Context-free Hypergraph Grammars

Node and Hyperedge Rewriting
with an Application to Petri Nets

von

Renate Klempien-Hinrichs

Dissertation

zur Erlangung des Grades einer
Doktorin der Ingenieurwissenschaften
– Dr.-Ing. –

Vorgelegt im Fachbereich 3 (Mathematik und Informatik)
der Universität Bremen
im Juli 2000

Die Deutsche Bibliothek – CIP-Einheitsaufnahme

Klempien-Hinrichs, Renate:
Context-free hypergraph grammars : node and hyperedge rewriting with
an application to Petri nets / Renate Klempien-Hinrichs. - Bremen :
R. Klempien-Hinrichs; Norderstedt : Books on demand, 2001
    Zugl.: Bremen, Univ., Diss., 2000
    ISBN 3-8311-2674-7

Herstellung: Books on Demand GmbH

Datum des Promotionskolloquiums: 22. September 2000

Gutachtende: Prof. Dr. Hans-Jörg Kreowski (Universität Bremen)
             Prof. Dr. Annegret Habel   (Universität Oldenburg)

In memoriam
Dr. med. Margot Klempien-Loelf

Acknowledgements

> There may be people who, in all serenity,
> find a topic, carry out research, and write a thesis.
> Suffice it to say that I am not one of them.

My first thanks go to Hans-Jörg Kreowski who, on the briefest of acquaintance, accepted to supervise my thesis. He enriched my chosen research topic with challenging questions, always gave good advice on my work in particular when I asked for it, and has greatly influenced my perception of science. Furthermore, I am thankful for his creativity when it came to finding funding for my work.

My second supervisor, Annegret Habel, convinced me that the research I had done in a couple of years could be shaped up to become a thesis. For quite a few times since then, she invited me to visit her for one or two days' working together, and always provided precise ideas on how to continue my thesis project. I would not have succeeded without her remarkable support and patience.

Michel Bauderon was in charge of me during the eight months I spent at the Université de Bordeaux I, France. He motivated me not only with his research ideas, but also with his positive reactions to what I made out of them.

Since my joining the Research Group Theoretical Computer Science at the Universität Bremen, I have enjoyed a friendly, supportive, and open exchange with my colleagues Frank Drewes, Sabine Kuske, and Detlef Plump. Frank, in particular, has had to rise manfully to numerous requests for advice, taught me how to use his *graph stzle*, gave me the raven to play with, and made me laugh even in the very bleak moments. Furthermore, my former colleagues Katrin Floegel and Angelika Hoppé were a source of strength in various discussions on the general difficulties associated with writing a thesis, and Carolina von Totth assisted me with some file format conversion.

While in Bordeaux, I have had the good fortune to collaborate with Hélène Jacquet. That cooperation has been turned into a pleasure by her capabilities to teach and to organise and by her friendship.

My parents Monika Adler and Heinrich Hinrichs and also my brother Heinrich Hinrichs have, on many occasions, relieved me of time-consuming duties or aided me therein. Your unconditional support is and has always been very important to me.

Finally, I gratefully acknowledge the financial support which I have received for the work reported here from the Universität Bremen, the EC TMR Network GET-GRATS (General Theory of Graph Transformation Systems) through the Université de Bordeaux I, and the ESPRIT Working Group APPLIGRAPH (Applications of Graph Transformation).

Contents

While the spells and enchantments which follow use commonly
available ingredients, if you have any allergies or sensitivities to
a particular ingredient, refrain from attempting that spell. Heed
all warnings and instructions on the products you use.

— Lexa Roséan —

Introduction

In order to allow humans to develop, modify, and also simply to understand highly complex systems, it is indispensable to have well-structured models of these systems available in various levels of detail or abstraction. Graphs, and more generally hypergraphs, provide adequate models for a wide variety of systems. In analogy to context-free string grammars, a derivation tree of a context-free hypergraph grammar represents the structure of the derived hypergraph in such a way that if the yield of the tree models some system, then each cut of the tree is a—more or less abstract—view on the same model. Moreover, it is easy to modify a model by replacing branches of derivation trees. The idea motivating the work reported in the present thesis is to combine the two most important context-free (hyper)graph grammar approaches, namely confluent node rewriting and hyperedge rewriting, into a framework where the strengths particular to the original approaches are maintained. For this, confluent node rewriting in hypergraphs is established as the most powerful context-free hypergraph grammar approach, and subsequently combined with the well-studied hyperedge-rewriting approach. Moreover, these methods to rewrite hypergraphs are shown to offer adequate formal frameworks for Petri net refinements.

Context-free Hypergraph Grammars

In most branches of science, graphs are acknowledged to provide adequate models for discrete systems, with the added bonus of a graphical representation which appeals to human intuition. This wide recognition is due to the fact that, as often pointed out in textbooks on graph theory such as [Har69, Che76], graphs represent binary relations and can therefore be used to model any system involving such a relation. There are, however, systems which involve relations of orders other than two. These systems can be modelled by *hypergraphs*: where in a graph an edge distinguishes a pair of nodes which is in the binary relation, a *hyperedge* in a hypergraph distinguishes a set [Ber73] or a sequence [Hab92a] of k nodes which is in some $k$ary relation. Examples for systems of this kind, as given by Habel [Hab92a], are functional expressions, Petri nets, flow diagrams, and chemical formulae in their structural representation. Further examples are electric circuits, entity-relationship diagrams, and hypertexts.

Most systems change in time: they may e.g. grow, shrink, change states, form subsystems, or unite with other systems. If graphs are chosen as system models, such changes translate into the—often rule-based—modification of graphs. This is the subject of graph grammars and graph transformation, an area which, initiated by the seminal papers [PR69, Sch70, EPS73], is well documented in the proceedings of several workshops [CER79, ENR83, ENRR87, EKR91, CM95, CEER96, EEKR00] and the three-volume *Handbook of Graph Grammars and Computing by Graph Transformation* [Roz97, EEKR99, EKMR99].

Graph grammars are a generalisation of Chomsky grammars from strings to graphs. In such a grammar, the application of a production consists of *choosing*, in the host graph, a subgraph to be rewritten which corresponds to the left-hand side of the production, *removing* that subgraph to yield the remainder of the host graph, *inserting* disjointly the right-hand side of the production, and *linking* the replacing graph with the remainder. Various graph grammar approaches exist, differing in particular in

- the type of graphs considered: simple or with parallel edges, directed or undirected, labelled (on nodes or edges or both) or not;

- the kind of subgraphs which can be rewritten: nodes, edges resp. hyperedges, handles (i.e. hyperedges together with their incident nodes), or arbitrary subgraphs;

- the way in which the replacing graph can be linked to the remainder of the original graph: connecting by inserting new edges, gluing by identifying nodes, or both;

- the mathematical framework used to describe the approach: set theory, category theory, or universal algebra.

For the generation of strings, and in particular strings which are programs in a programming language, context-free grammars have proved to be a particularly successful device. They are distinguished by the property that whenever a production is applied, exactly one nonterminal symbol is rewritten, and there is no application condition other than that this symbol constitutes the left-hand side of the production. For hypergraph grammars, the counterparts of these properties are that a primitive item, i.e. a node, a hyperedge, or a handle, can be rewritten in a derivation step, and that the left-hand side of the applied production consists of one nonterminal symbol, the label of the replaced nonterminal item. Thus, there are three different types of context-free graph resp. hypergraph grammars.

- Node-rewriting graph grammars are based on the connecting approach: nodes of the replacing graph can be linked by edges to neighbours of the rewritten node, as sketched in Figure 1. These grammars were introduced as a special case by Nagl [Nag76, Nag79]. The framework used there and in the majority of the literature (see [ER97] and the references therein) is set theory; universal algebra [Cou87] and category theory [Bau95a] are sometimes used, too.

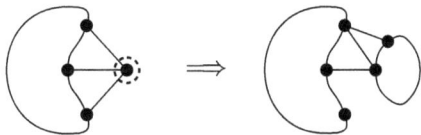

Figure 1: Rewriting the encircled node

- Hyperedge-rewriting graph and hypergraph grammars are based on the gluing approach: distinguished nodes of the replacing hypergraph are identified with the incident nodes of the rewritten hyperedge, as sketched in Figure 2. The idea was already introduced by Feder [Fed71] and Pavlidis [Pav72]; intensive research started with the approach by Bauderon and Courcelle [BC87], which is based on universal algebra, and the approach by Habel and Kreowski [HK87], which is based on category theory and usually presented in the framework of set theory. Further references can be found in [Hab92a, DHK97].

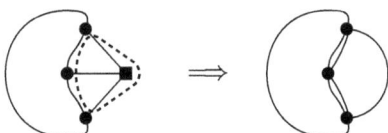

Figure 2: Rewriting the encircled hyperedge

- Similar to node-rewriting graph grammars, handle-rewriting hypergraph grammars are based on the connecting approach: nodes of the replacing hypergraph can be linked by hyperedges to neighbours of the nodes which

belonged to the rewritten handle, as sketched in Figure 3. These grammars
were introduced by Courcelle, Engelfriet, and Rozenberg [CER93] and treated
there in the framework of both set theory and universal algebra. A second
handle-rewriting approach was introduced by Kim and Jeong [KJ99].

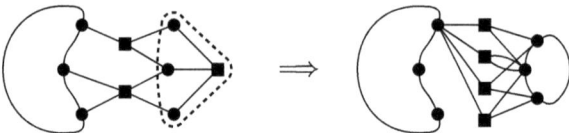

Figure 3: Rewriting the encircled handle

   Probably the most important reason for the success of context-free string grammars is that the 'structure' of a generated string can be described by means of a
syntax tree. Such a tree is also called derivation tree because it is a compact representation of a derivation; in fact, a derivation of the string can be constructed
from any traversal of the tree. In contrast to rewriting in strings, one does not
obtain context-free hypergraph grammars allowing to work with derivation trees
as described above just by restricting the size of parts which may be rewritten.
Following the abstract notion of context-freeness studied by Courcelle [Cou87], a
grammar is context-free only if it is associative and confluent, i.e. production applications can be evaluated in any order. Node-rewriting graph grammars are associative, but not necessarily confluent. Their confluent subclass was for the first time
considered in [Kau85]. Hyperedge-rewriting graph and hypergraph grammars are
naturally context-free. Handle-rewriting hypergraph grammars are made context-
free by considering only their separated subclass, i.e. two nonterminal handles may
not have any node in common.
   Given a hypergraph grammar which is context-free in the sense of [Cou87], a
derivation tree may be seen as a hierarchical hypergraph, and hence a hierarchical
representation of any system which is obtained as an interpretation of the hypergraph yielded by the tree. An example is provided by Petri net refinement, where
the refinement of a place or transition is described by the (context-free) replacement of an atomic item—a node or a hyperedge—in a hypergraph. Then one comes
to a notion of hierarchical net such as in e.g. [Feh93].

Contents and Structure of the Thesis

Rounding off the collection of context-free hypergraph grammar approaches, node
rewriting in hypergraphs is introduced and investigated in this thesis. In particular, a confluent subclass of these grammars is shown to be more powerful than

hyperedge rewriting and separated handle rewriting together, and confluence is proved to be decidable for the new approach. Moreover, a type of hypergraph grammars which combines node and hyperedge rewriting is studied, and Petri net refinement is established as an application. Finally, pullback rewriting is shown to provide a framework based on category theory for both node rewriting in hypergraphs and net refinement. Major parts of the work presented here are published in [Kle96, Kle99, Kle, HK98, JK00, Kle98].

Throughout the thesis, various forms of graphs and hypergraphs and context-free (hyper)graph grammar approaches are considered. To give the reader a point of reference, basic and central notions and notations which are recurring in several chapters are grouped together in Chapter 1.

In Chapter 2, node rewriting in hypergraphs is developed, generalising the well-known edNCE approach to node rewriting in graphs. The resulting hNCE approach behaves with respect to context-freeness exactly as the edNCE approach, i.e. an hNCE grammar is context-free if and only if it is confluent.

Chapter 3 contains results on the generative power of context-free hNCE grammars. The hNCE approach provides a highly flexible method to transform hyperedges connecting the node to be rewritten with its neighbourhood into hyperedges connecting the replacing hypergraph with that neighbourhood; in particular, the transformation need not preserve the rank of the hyperedges. Somewhat similarly, in hyperedge rewriting a hyperedge may be replaced with a hypergraph containing hyperedges of any rank. This entails that there is a proper hierarchy of hyperedge-rewriting languages depending on the rank of nonterminal hyperedges needed. For the languages generated by the so-called remote-free subclass of confluent hNCE grammars, however, there is no such hierarchy; in particular, the graph languages generated by this class are exactly the confluent edNCE languages. Moreover, the class of remote-free confluent hNCE languages promises to play the same important role for hypergraph languages as confluent edNCE languages do for graph languages. This is witnessed in particular by that class containing properly the union of languages generated by hyperedge rewriting or by separated handle rewriting, two incomparable hypergraph language classes. The chapter ends with two results on the minimal resp. maximal 'density' of confluent hNCE languages.

With the class of confluent hNCE languages as important as indicated by the results contained in Chapter 3, it is of interest to know whether a given hNCE grammar is confluent. Confluence is a property of the derivations, and while it is not difficult to verify its presence or absence for edNCE grammars, the situation is notably more complex for hNCE grammars. Thus, the decision algorithm presented in Chapter 4 is doubly exponential in the size of the input hNCE grammar.

In Chapter 5, a combination of node rewriting and hyperedge rewriting is developed, the so-called atom-replacement grammars. Associative (and confluent) atom-replacement languages are closed under hyperedge substitution with special atom-replacement languages. One can for example generate the set of all graphs, all tournaments, or all discrete graphs with n^2 nodes (for $n \in \mathbb{N}$); these languages cannot be generated by confluent node rewriting. However, the constructed atom-replacement grammars are in general not context-free.

The application of context-free hypergraph rewriting to Petri nets is studied in Chapter 6. While in the literature, net refinement has been related in an informal way to hyperedge rewriting, it turns out that confluent node rewriting is the only hypergraph-rewriting technique allowing to adequately model the transition refinement operation of [GG90]. Moreover, context-free hypergraph grammars provide an adequate framework for the generation and modification of specific classes of Petri net models, as witnessed by a hyperedge-rewriting grammar generating the structured workflow nets of [vdA97].

Whereas hyperedge rewriting has grown in the category-theoretical framework of the double-pushout approach by Ehrig [Ehr79], for node rewriting such a framework has been made available but quite recently in the form of the pullback approach by Bauderon [Bau95a, Bau96]. In Chapter 7, it is shown that hNCE rewriting can be translated into this approach. In addition, changing the underlying category from hypergraphs to nets allows to obtain a concise characterisation of the refinement method in [GG90].

In the chapter concluding the thesis, the results are assessed in a wider context, and topics for further research are pointed out.

1

Basic Concepts

This chapter establishes the basic vocabulary used in the present thesis. General mathematical notations are defined in Section 1.1. Section 1.2 fixes our notion of graphs and hypergraphs. Notations and concepts for rewriting systems and (context-free) grammars are given in Section 1.3.

1.1 Mathematical Notations

The set of natural numbers $\{0, 1, 2, 3, \ldots\}$ is denoted by $\mathbb{N}$, and $\mathbb{N}_+ = \mathbb{N} \smallsetminus \{0\}$. For $n \in \mathbb{N}_+$, the set $\{1, \ldots, n\}$ is denoted by $[n]$.

Let A be some set and A^* the set of finite sequences over A; the empty sequence is λ. For a sequence $w = a_1 \ldots a_n \in A^*$, $|w| = n$ denotes its length and $w(i) = a_i$ its ith entry. The canonical extension of a mapping $f \colon A \to B$, where A, B are some sets, to strings is $f^* \colon A^* \to B^*$ with $f^*(\lambda) = \lambda$ and $f^*(aw) = f(a)f^*(w)$ for $a \in A$ and $w \in A^*$.

The disjoint union of two sets A and B is denoted by $A \uplus B$; sometimes this notation is used just to stress that A and B are already disjoint. The powerset of a set A containing all subsets of A is denoted by $\mathcal{P}(A)$. The size of a set A is $\#A$.

1.2 Graphs and Hypergraphs

In this thesis, graphs and hypergraphs are simple, directed, and have labels on nodes as well as edges. An introduction to general graph theory can be found in e.g. [Har69, Ber91, Bol98, Die00]. The definitions in this section are oriented at [CER93] and [DHK97].

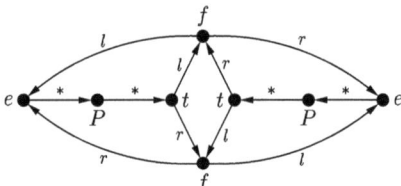

Figure 1.1: A graph

General assumption. Throughout the thesis, Σ denotes a finite set of symbols or *labels*; Σ may contain the symbol $*$ which means 'unlabelled' and is not drawn in pictures.

A (directed, node and edge labelled, loop-free) *graph* over Σ is a tuple $G = (V_G, E_G, lab_G)$ where V_G is a finite set of *nodes*, $E_G \subseteq \Sigma \times V_G \times V_G$ is a finite set of *edges* with $(\sigma, v, v) \notin E_G$ for all $\sigma \in \Sigma$ and $v \in V_G$, and $lab_G : V_G \to \Sigma$ is a mapping *labelling* a node v with $lab_G(v)$. The class of graphs over Σ is denoted by $\mathcal{G}_\Sigma$; just $\mathcal{G}$ is used if the particular set of symbols is not important. In a picture of a graph, a node v is represented by a circle $\bullet$ with $lab(v)$ written next to it and an edge (σ, u, v) by an arrow from u to v with σ written on it. Figure 1.1 shows a sample graph.

Hypergraphs generalise graphs in that a *hyperedge* may connect an arbitrary number of nodes instead of two distinct ones. Formally, a (directed, node and hyperedge labelled) *hypergraph* over Σ is a tuple $H = (V_H, E_H, lab_H)$ where V_H is a finite set of *nodes*, $E_H \subseteq \Sigma \times V_H^*$ is a finite set of *hyperedges*, and $lab : V_H \to \Sigma$ is a mapping *labelling* a node v with $lab_H(v)$. The class of hypergraphs over Σ is denoted by $\mathcal{H}_\Sigma$; just $\mathcal{H}$ is used if the particular set of symbols is not important. Each graph can be seen as a hypergraph by taking an edge (σ, v_1, v_2) as the hyperedge $(\sigma, v_1 v_2)$; in this sense, we have $\mathcal{G}_\Sigma \subseteq \mathcal{H}_\Sigma$. In pictures, a hyperedge $(\sigma, v_1 \dots v_n)$ is drawn as a square $\blacksquare$ with σ written next to it and linked to v_i by a line with i written on it. A hyperedge $\bullet\!-\!1\!-\!\blacksquare\!-\!2\!-\!\bullet$ which is also an edge may be drawn as an arrow $\bullet\!\longrightarrow\!\bullet$, too. Figure 1.2 shows a sample hypergraph.

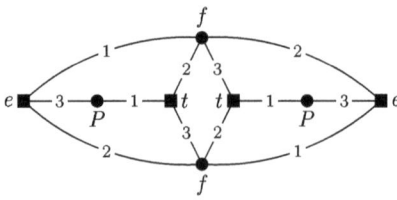

Figure 1.2: A hypergraph

Sometimes, we also consider hypergraphs which may have parallel hyperedges, that is more than one hyperedge with the same label and the same sequence of nodes. Formally, a *multiple hypergraph* over Σ is a tuple $M = (V_M, E_M, lab_M, att_M)$ where V_M and E_M are disjoint finite sets of nodes resp. hyperedges, $lab_M \colon V_M \cup E_M \to \Sigma$ assigns a label to nodes and hyperedges, and $att_M \colon E_M \to V_M^*$ assigns a sequence $att_M(e)$ of *attachment nodes* to each hyperedge $e \in E_M$. Two distinct hyperedges $e, e' \in E_M$ are *parallel* if $lab_M(e) = lab_M(e')$ and $att_M(e) = att_M(e')$. The class of multiple hypergraphs over Σ is denoted by $\mathcal{M}_\Sigma$; just $\mathcal{M}$ is used if the particular set of symbols is not important. Multiple hypergraphs are drawn in the same way as ordinary (simple) hypergraphs.

Multiple hypergraphs generalise hypergraphs (and graphs) as follows: A hypergraph $H = (V_H, E_H, lab_H)$ is interpreted as the multiple hypergraph $M_H = (V_M, E_M, lab_M, att_M)$ with $V_M = V_H$, $E_M = E_H$, $lab_M(v) = lab_H(v)$ for all $v \in V_M$, and $lab_M((\sigma, v_1 \ldots v_n)) = \sigma$ and $att_M((\sigma, v_1 \ldots v_n)) = v_1 \ldots v_n$ for all $(\sigma, v_1 \ldots v_n) \in E_M$. In this sense, we have $\mathcal{H}_\Sigma \subseteq \mathcal{M}_\Sigma$, and all notions which are defined below for multiple hypergraphs have the corresponding meaning for hypergraphs and, by extension, for graphs.

Let $M = (V_M, E_M, lab_M, att_M)$ be a multiple hypergraph. For a hyperedge $e \in E_M$ with $att_M(e) = v_1 \ldots v_n$, we denote the ith *incident* node of e by $att_M(e, i) = v_i$ (where i is said to be a *tentacle* of e), the *rank* of e by $rank_M(e) = n$, and the set of nodes incident to e by $vset_M(e) = \{att_M(e, i) \mid i \in [rank_M(e)]\}$. For a node $v \in V_M$, the set of hyperedges incident to v is $eset_M(v) = \{e \in E_M \mid v \in vset_M(e)\}$. Two distinct nodes $u, v \in V_M$ are *adjacent* or *neighbours* if there is a hyperedge $e \in E_M$ with $\{u, v\} \subseteq vset_M(e)$. A node or hyperedge x with label $lab_M(x) = \sigma$ is also called σ-*labelled*.

Let M and M' be multiple hypergraphs. A *hypergraph isomorphism* $f \colon M \to M'$ is a pair $f = (f_V, f_E)$ of bijections $f_V \colon V_M \to V_{M'}$ and $f_E \colon E_M \to E_{M'}$ such that $lab_{M'}(f(x)) = lab_M(x)$ for all $x \in V_M \cup E_M$ and $att_{M'}(f(e)) = f^*(att_M(e))$ for all $e \in E_M$. If an isomorphism $f \colon M \to M'$ exists, then M and M' are *isomorphic*, denoted $M \cong M'$. For a (*concrete*) multiple hypergraph M, the class $[M] = \{M' \mid M \cong M'\}$ of multiple hypergraphs isomorphic to M is an *abstract multiple hypergraph*.

The set of abstract multiple hypergraphs (simple hypergraphs, graphs) is denoted by $[\mathcal{M}_\Sigma]$ ($[\mathcal{H}_\Sigma]$, $[\mathcal{G}_\Sigma]$). A subset of $[\mathcal{M}_\Sigma]$ ($[\mathcal{H}_\Sigma]$, $[\mathcal{G}_\Sigma]$) is called a *language* of multiple hypergraphs (simple hypergraphs, graphs); the set of these languages is denoted by $\mathcal{LM}_\Sigma$ ($\mathcal{LH}_\Sigma$, $\mathcal{LG}_\Sigma$).

A *subhypergraph* of a multiple hypergraph $M = (V_M, E_M, lab_M, att_M)$ is a multiple hypergraph $M' = (V'_M, E'_M, lab'_M, att'_M)$ such that $V_{M'} \subseteq V_M$, $E_{M'} \subseteq E_M$, and $lab_{M'}$ and $att_{M'}$ are the respective restrictions of lab_M and att_M to $V_{M'}$ and $E_{M'}$. If $E_{M'} = \{e \in E_M \mid vset_M(e) \subseteq V_{M'}\}$, then M' is the subhypergraph *induced* by $V_{M'}$, denoted $M|_{V_{M'}}$.

For a multiple hypergraph H over Σ, a set $X \subseteq V_H \stackrel{.}{\cup} E_H$ of its nodes and hyperedges, and a set $\Sigma' \subseteq \Sigma$ of labels, $X|_{\Sigma'} := \{x \in X \mid lab_H(x) \in \Sigma'\}$ denotes the set of nodes and hyperedges of X labelled in Σ'.

1.3   Rewriting Systems, Grammars, and Context-freeness

Rewriting systems as a computational framework for term rewriting can be found in [BN98]; the specialisation as generative mechanisms in the form of grammars is considered in the context of formal language theory, see e.g. [Sal73, HU79]. The presentation of rewriting and grammars here is closely related to that in e.g. [ER97], but, in order to avoid needless repetitions for each concrete rewriting mechanism considered in this thesis, deals with abstract objects in a way similar to [Cou87]. This reference also contains an axiomatic definition of context-freeness which is presented below in a slightly more general form.

Rewriting systems.   Let $\mathcal{O}$ be a class of objects such that for an object $O \in \mathcal{O}$ containing some part x for which an object $O' \in \mathcal{O}$ disjoint from O can be substituted, this substitution results in an object $O[x/O'] \in \mathcal{O}$. Given a suitable notion of *isomorphy* for the objects in $\mathcal{O}$ (e.g. by hypergraph isomorphisms), the isomorphy of two objects O, O' is denoted by $O \cong O'$, and an *abstract* object is a class $[O]$ of objects isomorphic to a *concrete* object $O \in \mathcal{O}$. The class of abstract objects is denoted by $[\mathcal{O}]$. A *rewriting rule* $r = (X ::= O)$ over $\mathcal{O}$ consists of a left-hand side X (also denoted by $lhs(r)$) to abstractly specify the part of an object which can be rewritten with this rule, and a right-hand side $O \in \mathcal{O}$ (also denoted by $rhs(r)$) which describes how such a part will be transformed. A rewriting rule r can be applied to an object O at a part x of O yielding an object O' if x fits the specification of $lhs(r)$, $rhs(r)$ is disjoint from O, and $O' = O[x/rhs(r)]$. Such a *rule application* is denoted $O \Rightarrow_{[x,r]} O'$ and also called a *rewriting step* or a *direct derivation*.

A *rewriting system* is a pair $(\mathcal{O}, \mathcal{R})$ where $\mathcal{O}$ is a class of objects and $\mathcal{R}$ a finite set of rewriting rules over $\mathcal{O}$. Usually, $\mathcal{O}$ will be clear from the context and not mentioned explicitly. Two rewriting rules r, r' are isomorphic if $lhs(r) = lhs(r')$ and $rhs(r)$ is isomorphic to $rhs(r')$; we will always assume that the rewriting rules in $\mathcal{R}$ are pairwise nonisomorphic. The (infinite) set of all rewriting rules which are isomorphic to a rewriting rule in $\mathcal{R}$ is denoted by $copy(\mathcal{R})$, and a rule in $copy(\mathcal{R})$ is also called a *rule copy* of $\mathcal{R}$. A sequence

$$O = O_0 \Rightarrow_{[x_1,r_1]} O_1 \Rightarrow_{[x_2,r_2]} \cdots \Rightarrow_{[x_n,r_n]} O_n = O'$$

of n direct derivations with $r_i \in copy(\mathcal{R})$ for $i \in [n]$ and $n \in \mathbb{N}$ is called a *derivation* in $\mathcal{R}$ of *length* n. It is *creative* if the objects O and $rhs(r_i)$, $i \in [n]$, are mutually disjoint. We will restrict ourselves to creative derivations.

A derivation as above may also be written $O \Rightarrow_{\mathcal{R}}^n O'$ if we are not interested in the details of the rule applications, and $O \Rightarrow_{\mathcal{R}}^* O'$ if the precise length of the derivation is not important either. Thus, $\Rightarrow_{\mathcal{R}}^*$ is the reflexive transitive closure of the relationship $\Rightarrow_{\mathcal{R}}$ on $\mathcal{O}$.

Let $O \Rightarrow_{\mathcal{R}}^* O'$ be a derivation, x a part of O, and x' a part of O'. Then x is the *ancestor* of x' in O, denoted $anc_O(x')$, if $x = x'$, or $O \Rightarrow_{[x,r]} O'$ with $r \in \mathcal{R}$ and x' is a part of $rhs(r)$, or $O \Rightarrow_{\mathcal{R}}^* O'' \Rightarrow_{\mathcal{R}} O'$ and $x = anc_O(anc_{O''}(x'))$. Conversely, the *descendants* of x in O' form the set

$$desc_{O'}(x) = \{x' \mid x' \text{ is a part of } O' \text{ and } x = anc_O(x')\}.$$

Grammars. Let Σ be a finite set of symbols. An object *labelled* over Σ is an object O which has a set of items I_O and a mapping $lab_O : I_O \to \Sigma$ assigning a *label* $lab_O(i)$ to each item $i \in I_O$. (For example, a hypergraph H over Σ is an object labelled over Σ, where $I_O = V_H \cup E_H$.) We write $\mathcal{O}_\Sigma$ for a class of objects labelled over Σ.

Let $\mathcal{O}_\Sigma$ be a class of labelled objects. A *grammar* is a tuple $Gr = (N, T, P, Z)$ where the *alphabets* $N, T \subseteq \Sigma$ are finite, disjoint sets of *nonterminal* and *terminal* symbols respectively, P is a finite set of rewriting rules or *productions* where the left-hand sides are symbols in N and the right-hand sides are objects in $\mathcal{O}_{N \cup T}$, and $Z \in \mathcal{O}_{N \cup T}$ is the *initial object* or *axiom* of the grammar. Usually, Z is some elementary object induced by a label $S \in N$. In this case, we also write $Gr = (N, T, P, S)$. The pair $(\mathcal{O}_\Sigma, P)$ is the rewriting system associated with Gr, and the rule copies of P are also called production copies. The *sentential forms* of Gr form the set $S(Gr) = \{O \in \mathcal{O}_{N \cup T} \mid Z \Rightarrow_P^* O\}$, and the set of (abstract) objects *generated* by Gr is $L(Gr) = \{[O] \in [\mathcal{O}_T] \mid O \in S(Gr)\}$. Two grammars Gr and Gr' are *equivalent* if $L(Gr) = L(Gr')$. The class of languages which can be generated with grammars of some type X is denoted by $\mathcal{L}(X)$.

Context-freeness. Let $Gr = (N, T, P, Z)$ be a grammar. For an object $O \in \mathcal{O}_\Sigma$, let $sites(O)$ denote the set of parts of O which can be rewritten with a production. The grammar Gr is called *context-free* if it satisfies the following three axioms:

- PRESERVATION AXIOM:
  For all objects $O \in \mathcal{O}_\Sigma$, all production copies $(X ::= O') \in copy(P)$ with O' disjoint from O, and all parts $x \in sites(O)$ fitting X:
  $$sites(O[x/O']) = (sites(O) \smallsetminus \{x\}) \cup sites(O').$$

- CONFLUENCE AXIOM:
  For all sentential forms $O \in S(Gr)$, all production copies $(X_1 ::= O_1)$, $(X_2 ::= O_2) \in copy(P)$ with O, O_1, and O_2 mutually disjoint, and all parts $x_1, x_2 \in sites(O)$ with x_1 fitting X_1 and x_2 fitting X_2:

  1. $x_1 \in sites(O[x_2/O_2])$ and $x_2 \in sites(O[x_1/O_1])$, and
  2. $O[x_1/O_1][x_2/O_2] = O[x_2/O_2][x_1/O_1]$.

- ASSOCIATIVITY AXIOM:
  For all objects $O \in \mathcal{O}_\Sigma$, all production copies $(X_1 ::= O_1)$, $(X_2 ::= O_2) \in copy(P)$ with O, O_1, and O_2 mutually disjoint, and all parts $x_1 \in sites(O)$ fitting X_1 and $x_2 \in sites(O_1)$ fitting X_2:

  1. $x_2 \in sites(O[x_1/O_1])$ and
  2. $O[x_1/O_1[x_2/O_2]] = O[x_1/O_1][x_2/O_2]$.

Note that in [Cou87, p. 147], an abstract grammar is required to satisfy the preservation axiom, so that in this reference confluence and associativity are only defined for this case. In order to consider the more general case of grammars not satisfying the preservation axiom (see Chapter 5), it is useful to extend the confluence resp. associativity axiom by adding the respective first condition above. A further difference to [Cou87] is that we do not require an order on the rewriting sites of an object.

Finally, note that the definition of confluence as above differs from that usually considered for reduction systems such as term rewriting systems.

2

Node Rewriting in Hypergraphs

When rewriting a node in a graph with a second graph, the idea is to proceed as with string rewriting: remove the node from the first graph and insert the second graph in its place. But, as Figure 2.1 illustrates, something has to be done with the edges incident to the node, and there should be a way to link the two graphs through edges.

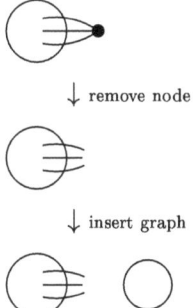

Figure 2.1: Node rewriting in a graph: What is to become of the edges?

The transformation of an edge incident to the rewritten node in the old graph into an edge linking the new graph to the rest of the old can be specified by augmenting graphs with sets of so-called connection instructions. The resulting graphs with embedding form a class of objects which allows for a notion of substitution in the sense of Section 1.3.

The edNCE approach (*NCE* stands for *N*ode-*C*ontrolled *E*mbedding, and *ed* for *e*dge-labelled, *d*irected graphs, it being understood that nodes have to be labelled in a node-rewriting approach) has become the standard approach to node rewriting in graphs (see [ER97, Eng97] and the references therein). The node- and edge-labelled directed graphs on which edNCE rewriting is defined are close to the class of hypergraphs on which hyperedge rewriting is defined. Therefore, the intuitive idea when developing a notion of node rewriting in hypergraphs is to generalise the edNCE mechanism of generating linking hyperedges from graphs to hypergraphs. The hNCE mechanism (h for $h$ypergraph) of generating linking hyperedges as presented in [Kle96] is such a generalisation.

The chapter is organised as follows. The edNCE approach is briefly recalled in Section 2.1 before the hypergraph-rewriting hNCE approach is introduced in Section 2.2. Examples for hNCE grammars are given in Section 2.3, and the question to which extent this approach is naturally context-free is discussed in Section 2.4. Finally, Section 2.5 contains some concluding remarks.

2.1   Node Rewriting in Graphs

The connection instructions with which a graph is augmented in the edNCE approach are of the form (ex/cr), where the existence part ex checks for edges having a certain format, which are then transformed into embedding edges according to the creation part cr. More precisely, $ex = (\alpha, \beta, d)$ describes an α-labelled edge between the node v to be replaced and a β-labelled neighbour of v which goes in the direction d (i.e. *in* or *out*, seen from v), and $cr = (\alpha', v', d')$ states that such an edge in the old graph gives rise to an α'-labelled edge between that neighbour and the node v' of the graph substituted for v, which goes in the direction d' (seen from v'). The effect of a connection instruction $(\alpha, \beta, in/\alpha', v', out)$ is sketched in Figure 2.2, where a graph containing a node v' is substituted for the node v.

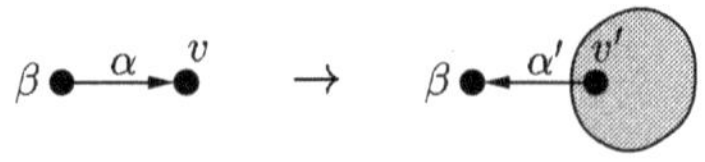

Figure 2.2: The effect of the connection instruction $(\alpha, \beta, in/\alpha', v', out)$

2.1 Definition (graph with embedding)
Let Σ be a finite set of symbols. A *graph with embedding* over Σ is a pair (G, C) where G is a graph over Σ and

$$C \subseteq (\Sigma \times \Sigma \times \{in, out\}) \times (\Sigma \times V_G \times \{in, out\})$$

is the *connection relation*. A *connection instruction* $coin = (ex/cr) \in C$ consists of an *existence part* $ex = (\alpha, \beta, d)$ and a *creation part* $cr = (\alpha', v', d')$. The set of all graphs with embedding over Σ is denoted by $\mathcal{GE}_\Sigma$.

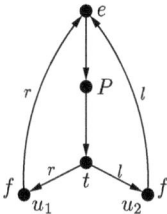

Figure 2.3: The graph G_2

A graph with embedding can be substituted for a node in another graph.

2.2 Example (node rewriting in a graph)

In the graph G_2 of Figure 2.3, let u_1 (resp. u_2) be the left (resp. right) f-labelled node and let C_2 be the connection relation $\{(l, e, out/l, u_1, out), (l, t, in/l, u_1, in), (r, t, in/r, u_2, in), (r, e, out/r, u_2, out)\}$. Then (G_2, C_2) is a graph with embedding.

Figure 2.4 illustrates the substitution of (G_2, C_2) for the upper f-labelled node v of the graph G_1 in Figure 1.1:

(1) REMOVE v together with all incident edges, yielding the remainder G_1^- of G_1;

(2) ADD G_2 to G_1^-; and

(3) CONNECT G_2 and G_1^- with edges according to the connection relation C_2.

For example, the existence of the l-labelled edge from v to the leftmost e-labelled node u together with the instruction $(l, e, out/l, u_1, out) \in C_2$ leads to the creation of the l-labelled edge from u_1 to u. ∎

Graphs with embedding are suitable objects for a notion of (node) substitution in the sense of Section 1.3. Therefore, the effect of a rule application on a connection instruction has to be considered. Let v be the node in a graph with embedding (G, C) to be rewritten. Then a connection instruction such as e.g. $(\alpha, \beta, in/\alpha'', v, out) \in C$ means that an edge such as the one on the left of Figure 2.5 is transformed into the one at the top when substituting (G, C) for the node u. If v is now rewritten with a graph (G', C') and C' contains e.g. a connection instruction $(\alpha'', \beta, out/\alpha', v', in)$, then the edge at the top is transformed into the one on the right. So, if v in (G, C) is rewritten with (G', C') yielding (G'', C''), then substituting (G'', C'') for u should transform the edge on the left directly into the one on the right, i.e. C'' should contain the instruction $(\alpha, \beta, in/\alpha', v', in)$. This relationship can also be seen by putting the two original connection instructions on top of each other to derive the third from them:

$$\frac{(\alpha, \beta, in/\, \alpha'', v'\, out\,) \qquad\qquad}{\qquad\qquad (\alpha'', \beta, out/\alpha', v', in)}$$
$$(\alpha, \beta, in \qquad / \qquad \alpha', v', in)$$

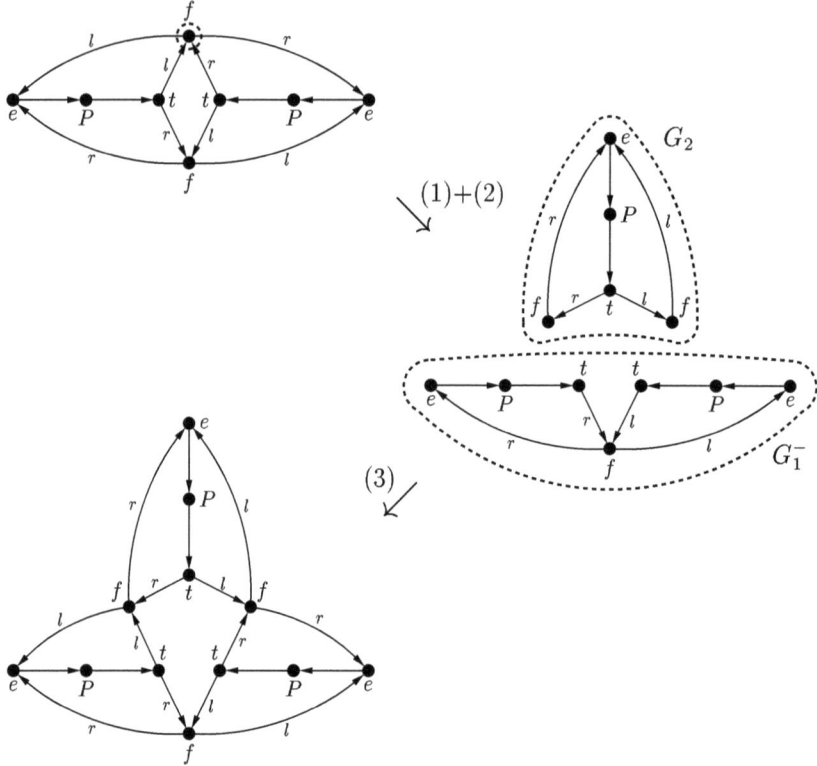

Figure 2.4: Substituting (G_2, C_2) for a node

Moreover, each connection instruction of C where v does not appear in the creation part has to belong to the resulting graph with embedding.

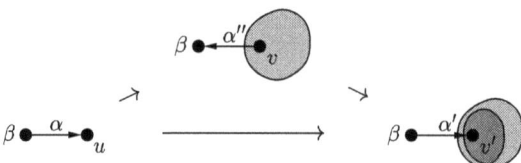

Figure 2.5: The effect rewriting v can have on
a connection instruction $(\alpha, \beta, in/\alpha'', v, out)$

2.3 Definition (edNCE rewriting)
Let (G_1, C_1), (G_2, C_2) be two disjoint graphs with embedding and v a node in G_1.
Then $(G_1, C_1)[v/(G_2, C_2)]$ is the graph with embedding (G_3, C_3) defined as follows,
where $G_i = (V_i, E_i, lab_i)$ for $i \in [3]$:

- $V_3 = (V_1 \smallsetminus \{v\}) \cup V_2$,

- $E_3 = \{e \in E_1 \mid v \notin vset_1(e)\}$
  $\cup E_2$
  $\cup \{(\alpha', u, v') \mid \exists \alpha \in \Sigma \colon (\alpha, u, v) \in E_1, (\alpha, lab_1(u), in/\alpha', v', in) \in C_2\}$
  $\cup \{(\alpha', v', u) \mid \exists \alpha \in \Sigma \colon (\alpha, u, v) \in E_1, (\alpha, lab_1(u), in/\alpha', v', out) \in C_2\}$
  $\cup \{(\alpha', u, v') \mid \exists \alpha \in \Sigma \colon (\alpha, v, u) \in E_1, (\alpha, lab_1(u), out/\alpha', v', in) \in C_2\}$
  $\cup \{(\alpha', v', u) \mid \exists \alpha \in \Sigma \colon (\alpha, v, u) \in E_1, (\alpha, lab_1(u), out/\alpha', v', out) \in C_2\}$,

- $lab_3(u) = \begin{cases} lab_1(u) & \text{if } u \in V_1 \smallsetminus \{v\}, \\ lab_2(u) & \text{if } u \in V_2, \end{cases}$

- $C_3 = \{(\alpha, \beta, d/\alpha', v', d') \in C_1 \mid v' \neq v\}$
  $\cup \{(\alpha, \beta, d/\alpha', v', d') \mid \exists \alpha'' \in \Sigma, d'' \in \{in, out\} \colon$
  $(\alpha, \beta, d/\alpha'', v, d'') \in C_1, (\alpha'', \beta, d''/\alpha', v', d') \in C_2\}$.

2.2   A Generalisation to Hypergraphs

Following the generalisation from edges to hyperedges, the connection instructions
with which a hypergraph is augmented are still of the form (ex/cr), but now the
existence and creation parts describe hyperedges. Therefore, $ex = (\alpha, x_1 \ldots x_m)$
consists of a label α for the label of the described hyperedge, and a sequence
$x_1 \ldots x_m$ over $\Sigma \uplus \{\Diamond\}$ for the attachment sequence of this hyperedge. More precisely, the rank of the hyperedge is m, $x_i = \Diamond$ if and only if the ith attachment
node is the node v to be replaced, and otherwise $x_i \in \Sigma$ is the label of the ith
attachment node. The creation part $cr = (\alpha', y_1 \ldots y_n)$ consists of a label α' for
the label of the hyperedge to be created and a sequence $y_1 \ldots y_n$ over the nodes of
the graph to be substituted plus the positions $i \in \mathbb{N}_+$ with $x_i \in \Sigma$ for the attachment sequence of this hyperedge. More precisely, the rank of the new hyperedge
is n, the jth attachment node is y_j if y_j is a node, and the y_jth attachment node
of the old hyperedge if $y_j \in \mathbb{N}_+$. Clearly, this requires for every $y_j \in \mathbb{N}_+$ that the
rank of the old hyperedge is at least y_j, and its y_jth attachment node should not
be the replaced node. The effect of a connection instruction $(\alpha, \beta\Diamond\gamma/\alpha', u'3v'1)$
is sketched in Figure 2.6, where a hypergraph containing some nodes u' and v' is
substituted for the node v.

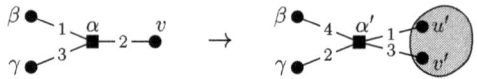

Figure 2.6: The effect of the connection instruction $(\alpha, \beta\Diamond\gamma/\alpha', u'3v'1)$

2.4 Definition (hypergraph with embedding)

Let Σ be a finite set of symbols. Then $EX = \Sigma \times (\Sigma \uplus \{\Diamond\})^*$ denotes the set of *existence parts* over Σ. For a hypergraph H, $CR_H = \Sigma \times (\mathbb{N}_+ \uplus V_H)^*$ denotes the set of *creation parts*. For an existence or creation part $z = (\alpha, x_1 \ldots x_n)$, we write $lab(z) = \alpha$, $rank(z) = n$, $z[1..n] = x_1 \ldots x_n$, and $z[i] = x_i$ for each $i \in [n]$. The set of *connection instructions* over Σ for a hypergraph H is

$$CI_H = \{(ex/cr) \in EX \times CR_H \mid \forall j \in [rank(cr)] \colon cr[j] \in \mathbb{N}_+ \Rightarrow \\ cr[j] \in [rank(ex)] \wedge ex[cr[j]] \in \Sigma\}.$$

A *hypergraph with embedding* over Σ is a pair (H, C) where H is a hypergraph over Σ and the *connection relation* $C \subseteq CI_H$ is a finite set of connection instructions over Σ for H. The set of all hypergraphs with embedding over Σ is denoted by $\mathcal{HE}_\Sigma$.

Two hypergraphs with embedding (H_1, C_1), $(H_2, C_2) \in \mathcal{HE}_\Sigma$ are isomorphic if there is a hypergraph isomorphism $f \colon H_1 \to H_2$ such that

$$C_2 = \{(ex/cr_2) \mid \exists (ex/cr_1) \in C_1 \colon \\ lab(cr_2) = lab(cr_1),\ rank(cr_2) = rank(cr_1), \\ \forall i \in [rank(cr_2)] \colon cr_2[i] = f(cr_1[i])\ \text{if}\ cr_1[i] \in V_{H_1},\ \text{and} \\ cr_2[i] = cr_1[i]\quad \text{if}\ cr_1[i] \in \mathbb{N}_+\}.$$

A hypergraph with embedding can be substituted for a node in another hypergraph.

2.5 Example (node rewriting in a hypergraph)

In the hypergraph H_2 of Figure 2.7, let u_1 (resp. u_2) be the upper (resp. lower) P-labelled node and u_3 (resp. u_4) be the upper (resp. lower) f-labelled node. Moreover, let C_2 be the connection relation consisting of the instructions $(t, \Diamond ff/t, u_1u_33)$, $(t, \Diamond ff/e, u_33u_1)$, $(e, ff\Diamond/t, u_21u_4)$, and $(e, ff\Diamond/e, 1u_4u_2)$. Then (H_2, C_2) is a hypergraph with embedding.

Figure 2.8 illustrates the substitution of (H_2, C_2) for the right P-labelled node v of the hypergraph H_1 in Figure 1.1:

(1) REMOVE v together with all incident hyperedges, yielding the remainder H_1^- of H_1;

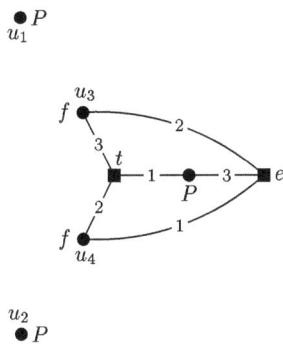

Figure 2.7: The hypergraph H_2

(2) ADD H_2 to H_1^-; and

(3) CONNECT H_2 and H_1^- with hyperedges according to the connection relation C_2.

For example, the existence of the t-labelled hyperedge linking v and its two f-labelled neighbours together with the instruction $(t, \Diamond ff/e, u_3 3u_1)$ in C_2 leads to the creation of the uppermost e-labelled hyperedge in the resulting hypergraph. ∎

Analogously to the edNCE approach, rewriting will be defined on hypergraphs with embedding, so the effect of a rule application on a connection instruction has to be considered. Let v be the node to be rewritten in a hypergraph with embedding (H, C). Then a connection instruction such as e.g. $(\alpha, \delta \Diamond \varepsilon / \beta, u'3v1) \in C$ means that a hyperedge such as the one on the left of Figure 2.9 is transformed into the one at the top when substituting (H, C) for the node u. If v is now rewritten with a hypergraph (H', C') and C' contains e.g. a connection instruction $(\beta, \zeta \varepsilon \Diamond \delta / \gamma, 4v'1)$, then the hyperedge at the top is transformed into the one on the right. So, if v in (H, C) is rewritten with (H', C') yielding (H'', C''), then substituting (H'', C'') for u should transform the hyperedge on the left directly into the one on the right, i.e. C'' should contain the instruction $(\alpha, \delta \Diamond \varepsilon / \gamma, 1v'u')$:

$$\frac{(\alpha, \delta \Diamond \varepsilon / \beta, u'3v1) \qquad (\beta, \zeta \varepsilon \Diamond \delta / \gamma, 4v'1)}{(\alpha, \delta \Diamond \varepsilon \quad / \quad \gamma, 1v'u')}$$

Moreover, each connection instruction of C where v does not appear in the creation part has to belong to the resulting hypergraph with embedding.

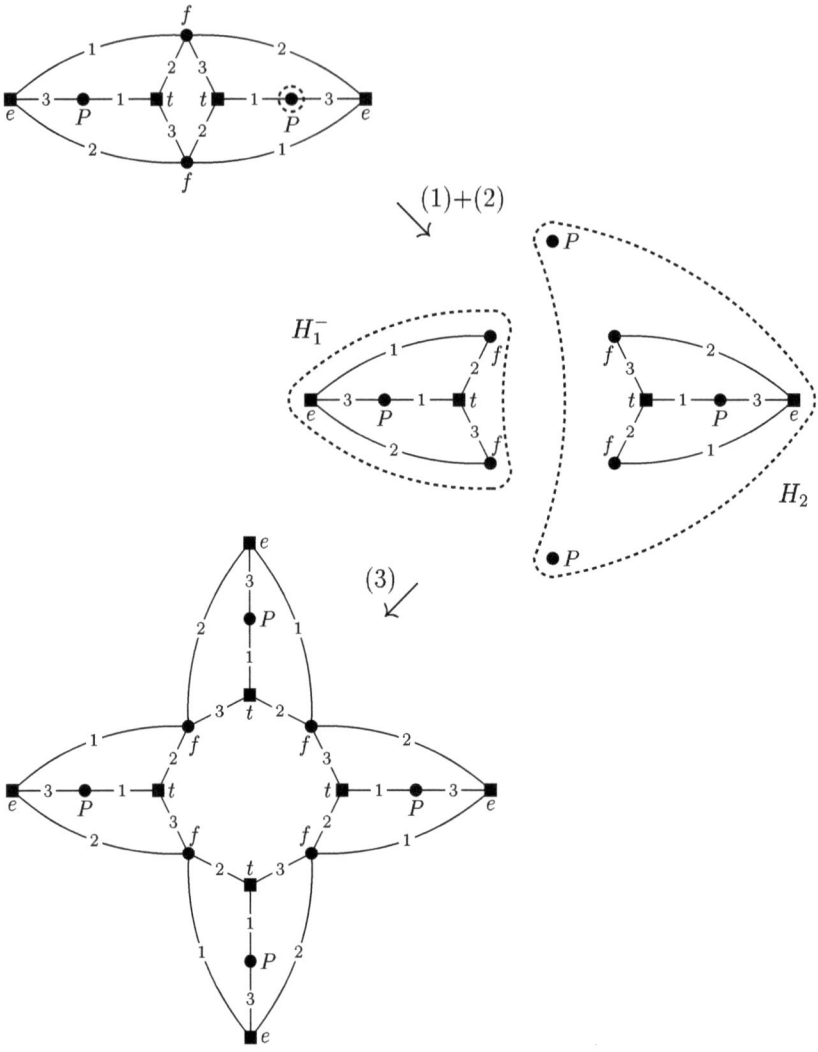

Figure 2.8: Substituting (H_2, C_2) for a node

2.6 Definition (hNCE rewriting)

Let (H_1, C_1), (H_2, C_2) be two disjoint hypergraphs with embedding and v a node in H_1. Then $(H_1, C_1)[v/(H_2, C_2)]$ is the hypergraph with embedding (H_3, C_3) defined as follows, where $H_i = (V_i, E_i, lab_i)$ for $i \in [3]$:

- $V_3 = (V_1 \setminus \{v\}) \cup V_2$,

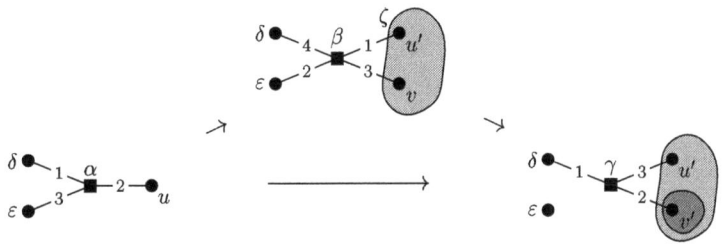

Figure 2.9: The effect rewriting v can have on
a connection instruction $(\alpha, \delta \Diamond \varepsilon / \beta, u'3v1)$

- $E_3 = \quad \{e \in E_1 \mid v \notin vset_1(e)\}$
  $\cup \ E_2$
  $\cup \ \{(lab(cr), v_1 \ldots v_{rank(cr)}) \mid \exists e \in E_1, \ (ex/cr) \in C_2:$
  $\qquad v \in vset_1(e), \ lab(ex) = lab_1(e), \ rank(ex) = rank_1(e),$
  $\qquad \forall i \in [rank(ex)]: \ (att_1(e, i) = v \wedge ex[i] = \Diamond) \vee$
  $\qquad\qquad\qquad\qquad\qquad (att_1(e, i) \neq v \wedge ex[i] = lab_1(att_1(e, i))),$
  $\qquad \forall j \in [rank(cr)]: \ (cr[j] \in V_2 \wedge v_j = cr[j]) \vee$
  $\qquad\qquad\qquad\qquad\qquad (cr[j] \in \mathbb{N}_+ \wedge v_j = att_1(e, cr[j]))\},$

- $lab_3(u) = \begin{cases} lab_1(u) & \text{if } u \in V_1 \setminus \{v\}, \\ lab_2(u) & \text{if } u \in V_2, \end{cases}$

- $C_3 = \quad \{(ex/cr) \in C_1 \mid \forall i \in [rank(cr)]: \ cr[i] \neq v\}$
  $\cup \ \{(ex/cr) \mid \exists (ex/cr_1) \in C_1, \ (ex_2/cr_2) \in C_2:$
  $\qquad \exists i \in [rank(cr_1)]: \ cr_1[i] = v,$
  $\qquad lab(ex_2) = lab(cr_1), \ rank(ex_2) = rank(cr_1),$
  $\qquad \forall i \in [rank(cr_1)]: \ (cr_1[i] = v \wedge ex_2[i] = \Diamond) \vee$
  $\qquad\qquad\qquad\qquad\qquad (cr_1[i] \in V_1 \setminus \{v\} \wedge ex_2[i] = lab_1(cr[i])) \vee$
  $\qquad\qquad\qquad\qquad\qquad (cr_1[i] \in \mathbb{N}_+ \wedge ex_2[i] = ex[cr_1[i]]),$
  $\qquad lab(cr) = lab(cr_2), \ rank(cr) = rank(cr_2),$
  $\qquad \forall j \in [rank(cr_2)]: \ (cr_2[j] \in V_2 \wedge cr[j] = cr_2[j]) \vee$
  $\qquad\qquad\qquad\qquad\qquad (cr_2[j] \in \mathbb{N}_+ \wedge cr[j] = cr_1[cr_2[j]])\}.$

On the basis of this notion of node rewriting in hypergraphs, it is useful to distinguish several types of connection instructions.

According to the definition of node rewriting, only connection instructions with at least one $\Diamond$ in the existence part (called local connection instructions) can give rise to an embedding hyperedge. While a meaningful interpretation of non-local connection instructions is conceivable (the existence part checks for hyperedges of a certain format which are anywhere in the hypergraph and not incident to the rewritten node), such a global operation would be in opposition to the focus of this thesis on context-free rewriting mechanisms. Therefore, we consider local connection instructions only.

General assumption. The connection relation of a hypergraph with embedding contains only local connection instructions.

A connection instruction is bridge-based if its existence part recognizes bridges, i.e. hyperedges which link the node to be replaced with at least one neighbour. For node rewriting in (loop-free) graphs, this is the only meaningful type of connection instructions.

A remote connection instruction creates hyperedges which are incident exclusively to neighbours of the replaced node, thus lying entirely in the context of that node.

During the embedding process, a link-preserving connection instruction can split a tentacle of a hyperedge recognized by its existence part in several tentacles (either to the same node, or to distinct nodes of the replacing hypergraph), but it may not erase a tentacle, i.e. lose the link represented by the tentacle. Moreover, for technical convenience the hyperedge is assumed to have exactly one tentacle to the replaced node.

A form-preserving connection instruction just redirects the tentacles of the recognized hyperedge from the replaced node to nodes of the replacing hypergraph, and it may change the hyperedge label in the process.

2.7 Definition (classification of connection instructions)
Let H be a hypergraph over Σ. A connection instruction $coin \in CI_H$ is called:

- *local* if there is $i \in [rank(ex)]$ with $ex[i] = \Diamond$;

- *bridge-based* if it is local and there is $i \in [rank(ex)]$ with $ex[i] \in \Sigma$;

- *remote* if $cr[j] \in \mathbb{N}_+$ for all $j \in [rank(cr)]$;

- *link-preserving* if it is not remote, there is exactly one $i \in [rank(ex)]$ with $ex[i] = \Diamond$, and for each $i \in [rank(ex)]$ with $ex[i] \in \Sigma$ there is $j \in [rank(cr)]$ with $cr[j] = i$;

- *form-preserving* if $rank(ex) = rank(cr)$ and for all $i \in [rank(ex)]$: $ex[i] \in \Sigma$ implies $cr[i] = i$, and $ex[i] = \Diamond$ implies $cr[i] \in V_H$.

The following translation of connection instructions allows to transfer the correspondence between graphs and certain hypergraphs to (hyper)graphs with embedding: A graph with embedding $(G, C) \in \mathcal{GE}_\Sigma$ corresponds to the hypergraph with embedding $(G, \tau(C)) \in \mathcal{HE}_\Sigma$, where $\tau(C) = \{\tau(coin) \mid coin \in C\}$ and for all $\alpha, \alpha', \beta \in \Sigma$, $v' \in V_G$:

$$
\begin{aligned}
\tau((\alpha, \beta, in/\alpha', v', in)) &= (\alpha, \beta\Diamond/\alpha', 1v'), \\
\tau((\alpha, \beta, in/\alpha', v', out)) &= (\alpha, \beta\Diamond/\alpha', v'1), \\
\tau((\alpha, \beta, out/\alpha', v', in)) &= (\alpha, \Diamond\beta/\alpha', 2v'), \\
\tau((\alpha, \beta, out/\alpha', v', out)) &= (\alpha, \Diamond\beta/\alpha', v'2).
\end{aligned}
$$

As we will not distinguish between (G, C) and $(G, \tau(C))$, we can say that $\mathcal{GE}_\Sigma \subseteq \mathcal{HE}_\Sigma$, and all notions which are defined for hypergraphs with embedding carry over to graphs with embedding. Note in particular that the connection instructions of a graph with embedding are bridge-based and link-preserving, but not form-preserving because the direction of an embedding edge may be changed during the embedding process.

A hypergraph $H \in \mathcal{H}_\Sigma$ will be considered to be the same as the hypergraph with embedding $(H, \emptyset) \in \mathcal{HE}_\Sigma$, so that we can say that $\mathcal{H}_\Sigma \subseteq \mathcal{HE}_\Sigma$ (and $\mathcal{G}_\Sigma \subseteq \mathcal{GE}_\Sigma$). The relationship between (hyper)graphs (with embedding) is summarised in the following diagram, where the arrows are the injections as discussed above.

$$
\begin{array}{ccc}
\mathcal{G}_\Sigma & \to & \mathcal{H}_\Sigma \\
\downarrow & & \downarrow \\
\mathcal{GE}_\Sigma & \to & \mathcal{HE}_\Sigma
\end{array}
$$

2.3 hNCE Grammars

With the notion of substitution in the hNCE approach from the previous section, hNCE grammars and the languages generated by them can be defined in a standard language-theoretic way.

2.8 Definition (hNCE grammar, derivation, language)
A *node-rewriting hypergraph grammar with neighbourhood controlled embedding* (hNCE grammar for short) is a tuple $NG = (N, T, P, S)$ where N and T are finite, disjoint sets of nonterminal and terminal symbols respectively, $P \subseteq N \times \mathcal{HE}_{NUT}$ is a finite set of productions, and $S \in N$ is the initial nonterminal of NG. A production $p = (X, (R, C)) \in P$ is usually written $X ::=_p (R, C)$ or simply $X ::= (R, C)$, its left-hand side is $lhs(p) = X$, and $rhs(p) = (R, C)$ its right-hand side. The hypergraphs of the right-hand sides of the productions in NG are the hypergraphs of NG. The hypergraph $\bullet S$ consisting of one S-labelled node is the axiom or initial hypergraph of NG.

Let $NG = (N, T, P, S)$ be an hNCE grammar. Moreover, let H and H' be hypergraphs with embedding over $N \cup T$, $v \in V_H$, and $p = (X ::= (R, C))$ (an isomorphic copy of) a production in P, with (R, C) disjoint from H. Then p can be applied to v in H if $lab_H(v) = X$. The application of p to v in H yields the hypergraph with embedding H', denoted $H \Rightarrow_{[v,p]} H'$, if $H' = H[v/(R, C)]$. A production application is also called a direct derivation or derivation step. If the exact information on v (and p) is not needed, it may also be written $H \Rightarrow_p H'$ ($H \Rightarrow_P H'$). For $n \in \mathbb{N}$, a derivation of length n consists of n consecutive derivation steps and is denoted by $H \Rightarrow_P^n H'$. If the length of a derivation is not important, we also write $H \Rightarrow_P^* H'$.

For an hNCE grammar $NG = (N, T, P, S)$, the set of sentential forms is $S(NG) = \{H \in \mathcal{H}_{N \cup T} \mid \bullet S \Rightarrow_P^* H\}$, and the generated hypergraph language is $L(NG) = \{[H] \in [\mathcal{H}_T] \mid H \in S(NG)\}$. The grammar is *graph-generating* if $L(NG) \subseteq [\mathcal{G}_T]$. The class of hypergraph languages which can be generated with hNCE grammars is denoted by $\mathcal{L}(\text{hNCE})$.

2.9 Example (hNCE grammars)

1. A *complete* graph is a graph in which each pair of nodes is linked by two edges in opposite direction. The set COMPLETE of all complete (a-labelled) graphs can be generated by the hNCE grammar $NG_1 = (N_1, T_1, P_1, S)$, where $N_1 = \{S\}$, $T_1 = \{a\}$, and P_1 contains

$$S ::= (\ a \overset{a}{\underset{u_1 \ a \ u_2}{\rightleftarrows}} S\ , C_1) \qquad \text{and} \qquad S ::= (\ \bullet a \atop u\ , C_2)$$
$$\quad p_1 \qquad\qquad\qquad\qquad\qquad\qquad p_2$$

   with $C_1 = \{(ex_{in}/a, 1u_1), (ex_{out}/a, u_12), (ex_{in}/a, 1u_2), (ex_{out}/a, u_22)\}$ and $C_2 = \{(ex_{in}/a, 1u), (ex_{out}/a, u2)\}$, where $ex_{in} = (a, a\Diamond)$ checks for an incoming and $ex_{out} = (a, \Diamond a)$ for an outgoing edge. A derivation in NG_1 is shown in Figure 2.10, where the unique terminal symbol a is omitted.

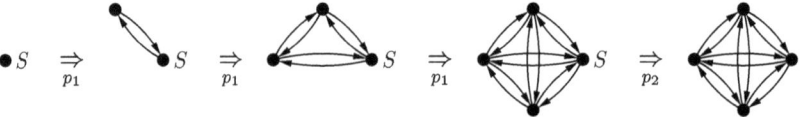

$\bullet S \underset{p_1}{\Rightarrow} \quad\quad S \underset{p_1}{\Rightarrow} \quad\quad S \underset{p_1}{\Rightarrow} \quad\quad S \underset{p_2}{\Rightarrow}$

Figure 2.10: A derivation in grammar NG_1 yielding the complete graph with four nodes

2. A *directed dotted tree* is a binary tree with an edge from each child node to its parent node and containing one additional node which has an incoming edge from each other node. The set DDTREES of all (edge-unlabelled) directed dotted trees (with b-labelled nodes, except for the distinguished node which

is d-labelled) can be generated by the hNCE grammar $NG_2 = (N_2, T_2, P_2, S)$, where $N_2 = \{S, X\}$, $T_2 = \{*, b, d\}$, and P_2 contains the productions

$$S \underset{p_1}{::=} (\, X \bullet \!\!\longrightarrow\!\! \bullet d \, , \, \emptyset)$$

$$X \underset{p_2}{::=} (\, \underset{u_1 \quad u_2}{\underset{X \bullet \quad \bullet X}{\overset{u_3}{\overset{\bullet b}{\bigwedge}}}} \, , \, \{(ex_d/cr_1), (ex_d/cr_2), (ex_d/cr_3), (ex_b/cr_3)\})$$

$$X \underset{p_3}{::=} (\, \underset{u_4}{\bullet} b \, , \, \{(ex_b/cr_4), (ex_d/cr_4)\})$$

where $ex_\alpha = (*, \Diamond\alpha)$ checks for an edge going out to an α-labelled neighbour ($\alpha \in \{b, d\}$), and $cr_i = (*, u_i 2)$ creates an edge going out from u_i to such a neighbour ($i \in [4]$). A derivation in NG_2 is shown in Figure 2.11.

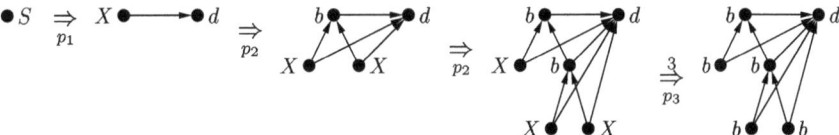

Figure 2.11: A derivation in grammar NG_2 yielding a directed dotted tree with $5 + 1$ nodes

3. For $k \in \mathbb{N}_+$, a k-hypertree is a 'tree' where a parent node is linked to its k children by a hyperedge of rank $k + 1$, with the $k + 1$st tentacle attached to the parent node. The set k-HYPERTREES of all k-hypertrees can be generated by the hNCE grammar $NG_3 = (N_3, T_3, P_3, S)$, where $N_3 = \{S\}$, $T_3 = \{*\}$, and P_3 contains the productions

$$S \underset{p_1}{::=} (\; \underset{S \bullet \, \cdots \, \bullet S}{\overset{u}{\overset{\bullet}{\underset{1 \quad k}{\overset{k+1}{\blacksquare}}}}} \; , \, C) \qquad \text{and} \qquad S \underset{p_2}{::=} (\, \bullet u \, , \, C)$$

with C containing all form-preserving connection instructions

$$(*, x_1 \ldots x_k */*, y_1 \ldots y_k (k+1))$$

over $\{S, *\}$ such that there is exactly one $i \in [k]$ with $x_i = \Diamond$, and $y_i = u$ for this i. A derivation in NG_3 (for $k = 2$) is shown in Figure 2.12.

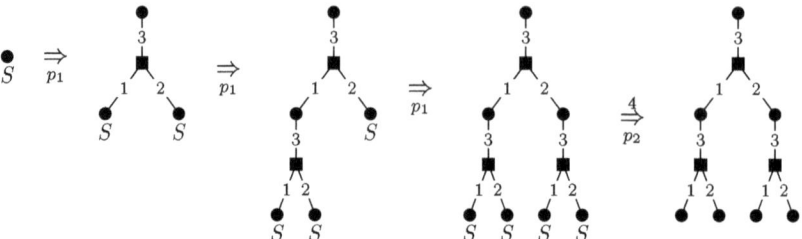

Figure 2.12: A derivation in grammar NG_3 yielding the balanced binary hypertree with three hyperedges

4. For $k \geq 2$, an *ordered k-tournament* is a hypergraph where the nodes can be arranged in a repetition-free sequence w such that the hyperedges are all hyperedges of rank k whose attachment sequence is a subsequence of w. The set k-TOURNAMENTS$_o$ of all ordered k-tournaments can be generated by an hNCE grammar. For $k = 3$, such a grammar is $NG_4 = (N_4, T_4, P_4, S)$, where $N_4 = \{S, X\}$, $T_4 = \{*\}$, and P_4 contains the productions

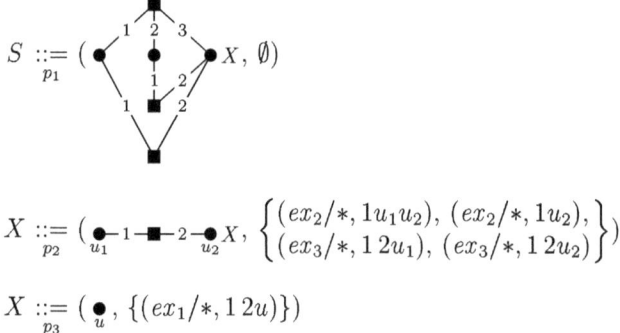

$$X ::= (\underset{p_2}{\bullet} \text{--}1\text{--}\blacksquare\text{--}2\text{--}\bullet X, \left\{ \begin{matrix} (ex_2/*, 1u_1u_2), \ (ex_2/*, 1u_2), \\ (ex_3/*, 1\,2u_1), \ (ex_3/*, 1\,2u_2) \end{matrix} \right\})$$

$$X ::= (\underset{p_3}{\bullet}, \{(ex_1/*, 1\,2u)\})$$

with $ex_2 = (*, *\Diamond)$ and $ex_3 = (*, **\Diamond)$, i.e. ex_i checks for a hyperedge of rank i gripping with its last tentacle to the replaced node. A derivation in NG_4 is shown in Figure 2.13, where tentacles are ordered from left to right. ∎

Given the translation of graphs with embedding into hypergraphs with embedding as discussed at the end of the previous section, edNCE grammars can now be defined as special hNCE grammars.

2.10 Definition (edNCE grammar)
An hNCE grammar $NG = (N, T, P, S)$ is a node-rewriting graph grammar with neighbourhood controlled embedding (edNCE grammar for short) if the right-hand side of every production in P is a graph with embedding over $N \cup T$. The class

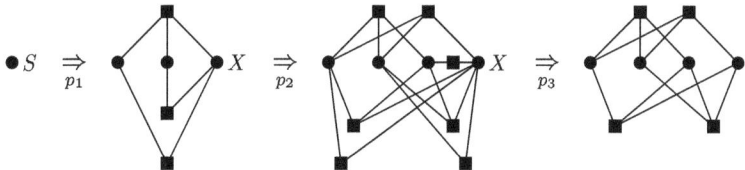

Figure 2.13: A derivation in grammar NG_4 yielding the ordered 3-tournament with four nodes (tentacles are ordered from left to right)

of graph languages which can be generated with edNCE grammars is denoted by $\mathcal{L}(\text{edNCE})$.

The grammars in Example 2.9.1, 2.9.2, and, for $k = 1$, 2.9.3 are edNCE grammars. Moreover, the language of ordered 2-tournaments can also be generated by an edNCE grammar.

The next lemma follows immediately from the definition of edNCE grammars.

2.11 Lemma
$\mathcal{L}(\text{edNCE}) \subsetneq \mathcal{L}(\text{hNCE})$.

Proof. Every edNCE grammar is an hNCE grammar generating a language of graphs, but no edNCE grammar can derive a hypergraph which is not a graph, such as a binary hypertree with at least one hyperedge.   □

While hNCE grammars generate hypergraphs and a hyperedge in a hypergraph may have more than just two incident nodes, a basic but important property of these grammars is that the hyperedge rank of the generated language is always bounded.

2.12 Definition (rank of a language or grammar)
Let A be a set of (concrete or abstract) hypergraphs. The rank of A, denoted $rank(A)$, is the maximal rank of a hyperedge occurring in a hypergraph of A, i.e. $rank(A) = \max\{rank_H(e) \mid H \in A, e \in E_H\}$.

Let $NG = (N, T, P, S)$ be an hNCE grammar. The rank of NG, denoted $rank(NG)$, is the maximal rank of a hyperedge or existence or creation part of a connection instruction occurring in NG, i.e.

$$rank(NG) = \max\left(\ \{rank_R(e) \mid (X ::= (R, C)) \in P, e \in E_R\}\right.$$
$$\cup\, \{rank(ex) \mid (X ::= (R, C)) \in P, (ex/cr) \in C\}$$
$$\left.\cup\, \{rank(cr) \mid (X ::= (R, C)) \in P, (ex/cr) \in C\}\ \right).$$

2.13 Lemma
For every hNCE grammar NG, $rank(S(NG))$ and $rank(L(NG))$ are bounded, and $rank(L(NG)) \leq rank(S(NG)) \leq rank(NG)$.

Proof. Let NG be an hNCE grammar. A hyperedge in a sentential form of NG either belongs to (a copy of) a hypergraph of NG or has been introduced by a connection instruction. There is only a finite number of hypergraphs of NG, each of them contains finitely many hyperedges, and all of them have a finite rank. Moreover, there is only a finite number of connection instructions in NG, and each creation part of such an instruction has finite rank. Thus, only finitely many hyperedges of distinct rank can occur in the sentential forms of NG, which implies that the maximal rank is bounded.

As $L(NG) \subseteq \{[H] \mid H \in S(NG)\}$, it is immediate that $rank(L(NG)) \le rank(S(NG))$. $\qquad\square$

2.4   Context-free hNCE Grammars

A grammar is context-free if rewriting a part of an object preserves the other rewriting sites of this object as well as those of the inserted object (preservation axiom), two direct derivations rewriting distinct sites of an object can be perfomed in any order without influencing the result (confluence axiom), and so can two direct derivations where the second rewrites a part of the object introduced by the first (associativity axiom). All hNCE grammars satisfy the preservation and associativity axioms, but not necessarily the confluence axiom.

Preservation of nonterminal nodes. Let $NG = (N, T, P, S)$ be an hNCE grammar and $sites(H, C) = lab_H^{-1}(N)$ denote the set of nonterminally labelled nodes in (H, C), for a hypergraph with embedding (H, C) over $N \cup T$. Then NG satisfies the preservation axiom if for all hypergraphs with embedding (H, C) over $N \cup T$, all production copies $(X ::= (R, C_R)) \in copy(P)$ with (R, C_R) disjoint from (H, C), and all nodes $v \in sites(H, C)$ with $lab_H(v) = X$ we have

$$sites((H, C)[x/(R, C_R)]) = (sites(H, C) \smallsetminus \{v\}) \cup sites(R, C_R).$$

2.14 Lemma
Every hNCE grammar satisfies the preservation axiom.

Proof. Let $NG = (N, T, P, S)$ be an hNCE grammar, $(H, C) \in \mathcal{HE}_{N \cup T}$, $v \in V_H$, and $(lab_H(v) ::= (R, C_R)) \in copy(P)$ with (R, C_R) disjoint from (H, C). Then the nodes of $(H, C)[v/(R, C_R)]$ are those of (R, C_R) and (H, C) minus v. Moreover, these nodes maintain their old labels, implying that each of them is nonterminal if and only if it has been so before. $\qquad\square$

Due to this lemma, the associativity and confluence axioms for hNCE grammars are a bit simpler than given in Section 1.3 for the general case.

Associativity. Let $NG = (N, T, P, S)$ be an hNCE grammar. Then NG satisfies the associativity axiom if for all hypergraphs with embedding (H, C) over $N \cup T$, all production copies $(X_1 ::= (R_1, C_1)), (X_2 ::= (R_2, C_2)) \in copy(P)$ such that (H, C), (R_1, C_1), and (R_2, C_2) are mutually disjoint, and all nodes $v_1 \in sites(H, C)$ with $lab_H(v_1) = X_1$ and $v_2 \in sites(R_1, C_1)$ with $lab_{R_1}(v_2) = X_2$, we have

$$(H, C)[v_1/(R_1, C_1)[v_2/(R_2, C_2)]] = (H, C)[v_1/(R_1, C_1)][v_2/(R_2, C_2)].$$

2.15 Lemma
Every hNCE grammar satisfies the associativity axiom.

Proof. Let $NG - (N, T, P, S)$ be an hNCE grammar, $(H, C) \in \mathcal{HE}_{NUT}$, $(X_1 ::- (R_1, C_1)), (X_2 ::= (R_2, C_2)) \in copy(P)$ such that (H, C), (R_1, C_1), and (R_2, C_2) are mutually disjoint, $v_1 \in sites(H, C)$ with $lab_H(v_1) = X_1$, and $v_2 \in sites(R_1, C_1)$ with $lab_{R_1}(v_2) = X_2$. Moreover, let

$$(H', C') = (H, C)[v_1/(R_1, C_1)][v_2/(R_2, C_2)]$$

and

$$(H'', C'') = (H, C)[v_1/(R_1, C_1)[v_2/(R_2, C_2)]].$$

It is straightforward to see that the nodes of H' are those of H'' and carry the same label in both cases. Similarly, the hyperedges of H, R_1, and R_2 which are not incident to either v_1 or v_2 occur both in H' and in H''. As each connection instruction in C_1 where v_2 is not present in the creation part also belongs to the connection relation of $(R_1, C_1)[v_2/(R_2, C_2)]$, the hyperedges which it generates appear in H' as well as in H''. A hyperedge of H which is transformed first by a connection instruction $coin_1 \in C_1$, then by a connection instruction $coin_2 \in C_2$ into a hyperedge of H', leads to the same hyperedge in H'' by the connection instruction generated from $coin_1$ through $coin_2$, and vice versa (cf. the discussion after Example 2.5). No hyperedges other than these belong to either H' or H'', so $H' = H''$. The arguments that $C' = C''$ are analogous to those concerning the hyperedges. $\square$

Confluence. Let $NG = (N, T, P, S)$ be an hNCE grammar. Then NG satisfies the confluence axiom if for all sentential forms $H \in S(NG)$, all production copies $(X_1 ::= (R_1, C_1)), (X_2 ::= (R_2, C_2)) \in copy(P)$ such that H, (R_1, C_1), and (R_2, C_2) are mutually disjoint, and all $v_1, v_2 \in sites(H)$ with $lab_H(v_1) = X_1$ and $lab_H(v_2) = X_2$, we have

$$H[v_1/(R_1, C_1)][v_2/(R_2, C_2)] = H[v_2/(R_2, C_2)][v_1/(R_1, C_1)].$$

2.16 Remark

In general, hNCE grammars do not satisfy the confluence axiom. This is already true for edNCE grammars, as the following example shows.

Consider an edNCE grammar NG which contains productions p_1, p_2 as follows:

- $p_1 = (A ::= (\, w \bullet a \,, \{(c, \Diamond B/c, w2), (c, \Diamond b/c, w2)\}))$ and

- $p_2 = (B ::= (\, {}^{x \bullet b}_{y \bullet b} \,, \{(c, a\Diamond/c, 1x), (c, A\Diamond/c, 1y)\})).$

Moreover, let NG have a sentential form consisting of a c-labelled edge from an A-labelled node u to a B-labelled node v. Then the derivations shown in Figure 2.14 are possible.

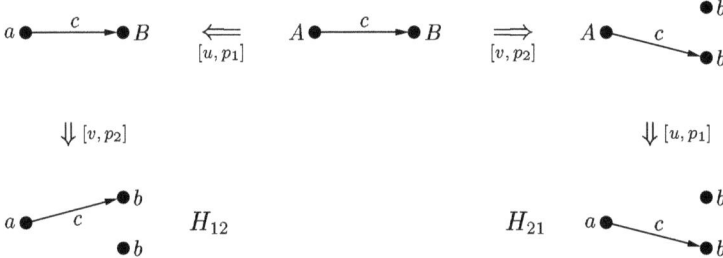

Figure 2.14: An example for non-confluence

While H_{12} and H_{21} are isomorphic, they are not equal, and thus NG does not satisfy the confluence axiom.

Context-freeness. As all hNCE grammars preserve nonterminal nodes and are associative, the notions of confluence and context-freeness can be used indifferently.

2.17 Definition (context-free hNCE grammar)

An hNCE grammar is *context-free* (a C-hNCE grammar for short) if it satisfies the confluence axiom. The class of hypergraph languages which can be generated with C-hNCE grammars is denoted by $\mathcal{L}$(C-hNCE); the class of graph languages which can be generated with context-free edNCE grammars (C-edNCE grammars for short) is denoted by $\mathcal{L}$(C-edNCE).

We immediately obtain:

2.18 Lemma

$\mathcal{L}$(C-edNCE) $\subsetneq$ $\mathcal{L}$(C-hNCE).

Proof. Analogous to the proof of Lemma 2.11. □

As a type of node-rewriting graph grammars which are naturally context-free, boundary grammars have been introduced by Rozenberg and Welzl [RW86]. The adaptation to hNCE grammars reads as follows.

2.19 Definition (boundary hNCE grammar)
Let $NG = (N, T, P, S)$ be an hNCE grammar. A hypergraph over $N \cup T$ is *boundary* if each node with label in N has only neighbours with labels in T (and thus a boundary of nodes which cannot be rewritten). The grammar NG is boundary (a B-hNCE grammar for short) if for each production $X ::= (R, C)$ in P, R is boundary and C contains only connection instructions (ex/cr) where $ex[1..rank(ex)] \in (T \cup \{\Diamond\})^*$ and at most one node of R occurring in $cr[1..rank(cr)]$ is labelled in N.

The grammars in Example 2.9.1, 2.9.2 and 2.9.4 are boundary.

An even simpler type of context-free grammars are the linear ones, investigated e.g. by Engelfriet and Leih [EL89].

2.20 Definition (linear hNCE grammar)
An hNCE grammar $NG = (N, T, P, S)$ is *linear* (an L-hNCE grammar for short) if for each production $(X ::= (R, C)) \in P$, R contains at most one node which is labelled in N.

The grammars in Example 2.9.1 and 2.9.4 are linear.

The following relationship holds between context-free, boundary, and linear hNCE grammars.

2.21 Lemma
Every boundary hNCE grammar is confluent. Every linear hNCE grammar has a boundary normal form.

Proof. It is immediate that all sentential forms of a boundary hNCE grammar are boundary themselves. As an hNCE grammar can only be non-confluent if there is a sentential form containing a hyperedge which is transformed to a different embedding hyperedge depending on the order in which two of its incident (nonterminal) nodes are rewritten (for a more detailed discussion see Chapter 4, in particular Lemma 4.2), the boundary condition guarantees confluence.

All sentential forms of a linear hNCE grammar contain at most one nonterminally labelled node. Thus, all connection instructions (ex/cr) with $ex[i] \in N$ for some $i \in [rank(ex)]$ can be deleted without changing the generated language, yielding a boundary (and therefore confluent) grammar. □

Linear, boundary, and confluent edNCE languages form a proper hierarchy, see the synopsis in [ER97, p. 59]. Whether the same holds for their respective hNCE counterparts is a matter of future research.

2.5   Concluding Remarks

The hNCE formalism as a set-theoretic formalisation of node rewriting in hypergraphs was introduced in [Kle96]. A category theoretical framework for node rewriting in graphs, called pullback rewriting, was developed at the same time in [Bau96]
and later generalised to hypergraphs in [BJ01b]. The basic idea of this approach
is to interpret a pullback square

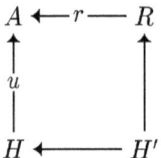

of (hyper)graph morphisms u, r as the application of a rule r to an unknown part
of H which is described by u, yielding the (hyper)graph H'. Pullback rewriting
is discussed in detail in Chapter 7, where a simulation of hNCE rewriting in this
approach is given.

3

The Generative Power
of Context-free hNCE Grammars

In the literature, a number of concepts can be found to generate sets of graphs
or hypergraphs in a recursive and context-free way. Context-freeness can only be
expected if a notion of graph or hypergraph transformation is used where small
items are replaced with larger graphs resp. hypergraphs. So far, confluent node
rewriting, hyperedge rewriting, and separated handle rewriting approaches have
been proposed, among them in particular:

- C-edNCE grammars (confluent node rewriting in graphs first studied by Kaul
  [Kau85]),

- HR grammars (hyperedge rewriting in hypergraphs introduced independently
  by Bauderon and Courcelle [BC87] and Habel and Kreowski [HK87]),

- S-HH grammars (separated handle rewriting in hypergraphs investigated by
  Courcelle, Engelfriet and Rozenberg [CER93]), and

- C-hNCE grammars (confluent node rewriting in hypergraphs presented by
  Klempien-Hinrichs [Kle96]).

While these approaches differ of course syntactically, it is important to know where
there are semantic differences, if any. Such a comparison can be done with respect
to their graph- or hypergraph-generating power.

The class of graph languages which is defined by C-edNCE rewriting coincides
with that defined by S-HH rewriting [CER93, Theorem 4.4] and contains properly
that defined by HR rewriting [ER90]. Considering that for HR grammars, Theorem 2.7 in [Hab92a, Chapter V] states that their graph- and hypergraph-generating

power increases with the maximal rank of the hyperedges occurring in a grammar, a similar result might be expected for C-hNCE grammars, which would in turn mean that C-hNCE grammars can generate more graph languages than C-edNCE grammars. However, there is no such hierarchy theorem for the so-called remote-free subclass of C-hNCE grammars, and their graph-generating power coincides with that of C-edNCE grammars, too [Kle].

The classes of hypergraph languages defined by HR resp. S-HH rewriting are incomparable, see [CER93]. In this reference, the question for a natural context-free class containing both HR and S-HH languages is put. C-hNCE languages, even remote-free C-hNCE languages, form such a class; moreover, there is a remote-free C-hNCE grammar whose generated language is neither an HR nor an S-HH language [Kle99].

While C-hNCE rewriting proves to be the most powerful known context-free hypergraph-generating technique, the question arises which sets of (hyper)graphs cannot be specified in this way. Clearly, all results known for C-edNCE grammars also hold for the graph-generating power of remote-free C-hNCE grammars. Moreover, similar results can be shown for the hypergraph-generating power of remote-free C-hNCE grammars, and for the generative power of general C-hNCE grammars.

In Section 3.1, the impact of the maximal rank of C-hNCE grammars on their generative power is investigated, and the graph languages generated by remote-free C-hNCE grammars are shown to be C-edNCE languages. Section 3.2 contains a simulation of HR grammars by remote-free C-hNCE grammars, and Section 3.3 a simulation of S-HH grammars. First results on limits to the generative power of C-hNCE grammars are presented in Section 3.4.

3.1   Normal Forms for C-hNCE Grammars

The hNCE approach is a canonical generalisation of the edNCE approach. Thus it is a natural question whether using hyperedges for the connection process enhances the expressive power of the formalism for the generation of graph languages. In this section, the problem is investigated for context-free grammars. It turns out that if remote connection instructions, i.e. those which produce hyperedges only incident to nodes in the host hypergraph, are disregarded, then the class of C-hNCE graph languages coincides with the class of C-edNCE languages [Kle].

The first main result of this section states that whenever a hypergraph language of rank k can be generated with a remote-free C-hNCE grammar, then an equivalent grammar can be constructed where only hyperedges of rank up to k occur. Subsequently, this result is employed to show that every remote-free C-hNCE grammar which generates a graph language has a C-edNCE normal form.

3.1 Definition (remote-free hNCE grammar)

Let NG be an hNCE grammar with production set P. A production $X ::= (H, C)$ in P is *remote-free* if C does not contain a remote connection instruction. The grammar NG is remote-free if all productions in P are. The class of remote-free hNCE grammars is denoted by $hNCE_{rf}$.

In an $hNCE_{rf}$ grammar, a hyperedge of rank greater than 0 occurring in some sentential form is generated as soon as its incident nodes are.

3.2 Lemma

Let NG be a remote-free hNCE grammar with production set P. For all sentential forms $H, H' \in S(NG)$ and each hyperedge $e \in E_{H'}$, if $H \Rightarrow^*_P H'$ and $\emptyset \neq vset_{H'}(e) \subseteq V_H$ then $e \in E_H$.

Proof. Suppose that $e \notin E_H$. Then there must be a hyperedge $e' \in E_H$ with $vset_{H'}(e) \subseteq vset_H(e')$ and the replacement of a node v incident to e', but not belonging to $vset_{H'}(e)$, is necessary to generate e as an embedding hyperedge. However, the connection instruction used to generate e in this case must be remote, a contradiction. □

The following result states that when defining a hypergraph language of rank k (for some $k \in \mathbb{N}$) with a $C\text{-}hNCE_{rf}$ grammar, only hyperedges of rank at most k need to be specified in the grammar. This is in contrast to HR grammars, where increasing the maximal rank of hyperedges leads to a proper increase of generative power [Hab92a, Chapter V, Theorem 2.7].

3.3 Theorem (rank theorem)

Let $k \in \mathbb{N}$. For every $C\text{-}hNCE_{rf}$ grammar generating a language of rank k, an equivalent $C\text{-}hNCE_{rf}$ grammar of rank k can be constructed.

Outline of the proof. We will first show that every $C\text{-}hNCE_{rf}$ grammar has a link-preserving normal form, which implies that the rank of hyperedges cannot decrease during embedding. A link-preserving C-hNCE grammar generating a language of rank k may still rely on nonterminally labelled hyperedges of rank greater than k to exclude sentential forms with terminal nodes only from the generated language, but a second normal-form theorem states that these so-called blocking hyperedges can be removed. The grammar which results from these normal-form constructions can then be transformed into a grammar of rank k. □

First we consider the link-preserving normal form of $C\text{-}hNCE_{rf}$ grammars, where the rank of a hyperedge cannot decrease during the embedding process.

3.4 Definition (link-preserving hNCE grammar)

Let NG be an hNCE grammar with production set P. A production $X ::= (H, C)$ in P is *link-preserving* if all connection instructions in C are. The grammar NG

is link-preserving if all productions in P are. The class of link-preserving hNCE grammars is denoted by hNCE$_{lp}$.

3.5 Theorem (link-preserving normal form)

For every C-hNCE$_{rf}$ grammar an equivalent link-preserving C-hNCE$_{rf}$ grammar can be constructed.

Proof. Let $NG = (N, T, P, S)$ be a C-hNCE$_{rf}$ grammar. A hyperedge is introduced into a sentential form of NG either as part of a right-hand side of a production or by a (succession of) connection instruction(s) transforming such an original hyperedge. Construct an hNCE grammar $NG' = (N', T, P', S)$ using the following idea: Replace each original hyperedge e which is incident to some nonterminal node with a set of *sub-hyperedges* each of which is incident to a distinct subset of the incident nodes of e (including at least one nonterminal node) and has sufficient information on the shape of e stored in its label.

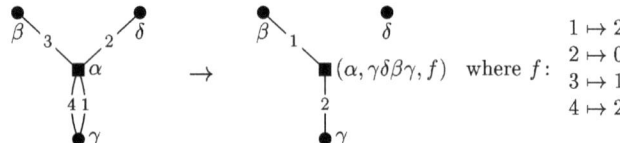

Figure 3.1: An example for the simulation of the hyperedge on the left

For the formal definition, let H and H' be two hypergraphs over $N \cup T$ with the same node set. A hyperedge $e' \in E_{H'}$ is a *sub-hyperedge* of a hyperedge $e \in E_H$ if $vset_{H'}(e') \subseteq vset_H(e)$ with $lab_{H'}(vset_{H'}(e')) \cap N \neq \emptyset$, the nodes in $att_{H'}(e')$ are pairwise distinct, and $lab_{H'}(e') = (lab_H(e), lab_H(att_H(e)), f)$ with $f: [rank_H(e)] \to [rank_{H'}(e')] \cup \{0\}$ such that $f(i) = j$ if $att_H(e, i) = att_{H'}(e', j)$ and $f(i) = 0$ otherwise.

An example is shown in Figure 3.1, where β or γ is assumed to be nonterminal and the hyperedge e' on the right is one of the sub-hyperedges of the hyperedge e on the left. The label of e' consists of the label of e, the labels of the attachment nodes of e, and finally a mapping $f: [rank(e)] \to [rank(e')] \cup \{0\}$ which maps a tentacle of e to the corresponding tentacle of e' (resp. to 0 if there is no such tentacle). Assuming that during some derivation, the hyperedge e evolves into a hyperedge $\bar{e}$ incident only to terminal descendants u, v of the β- and γ-labelled nodes, the confluence of the grammar implies that the derivation steps can be reordered such that u and v are generated before the δ-labelled node w is rewritten (if w is rewritten at all). Then the remote-freeness of the grammar implies that rewriting w does not influence the generation of $\bar{e}$, i.e. $\bar{e}$ is generated as soon as u and v are (Lemma 3.2). Thus, the incidences and the label of e' contain sufficient information to simulate the evolution from e to $\bar{e}$ by an evolution from e' to $\bar{e}$.

Formal construction of NG. Let $N' = N \uplus N_{info}$, where N_{info} contains the new nonterminal hyperedge labels and is defined as follows:

$$N_{info} = \{(\alpha, \beta_1 \ldots \beta_n, f) \mid n \in [rank(NG)],\ \alpha, \beta_1, \ldots, \beta_n \in N \cup T,$$
$$\exists i \in [n]\colon \beta_i \in N \wedge f(i) \neq 0,\ \exists k \in [n]\colon f\colon [n] \to [k] \cup \{0\},$$
$$\forall i, j \in [n]\colon f(i) = f(j) \Rightarrow \beta_i = \beta_j,\ \forall i \in [k]\colon f^{-1}(i) \neq \emptyset\}.$$

Let $P' = \{\lambda(p) \mid p \in P\}$, where for each production $p = (X ::= (H, C))$ in P and $H = (V_H, E_H, lab_H)$, the production $\lambda(p) = (X ::= (H', C'))$ with $H' = (V_{H'}, E_{H'}, lab_{H'})$ is constructed as follows:

- $V_{H'} = V_H$ and $lab_{H'}(v) = lab_H(v)$ for all $v \in V_{H'}$,

- $E_{H'} = \{e \in E_H \mid lab_H(att_H(e)) \in T^*\} \cup$
  $\{e' \mid \exists e \in E_H \colon lab_H(vset_H(e)) \cap N \neq \emptyset,\ e'$ is a sub-hyperedge of $e\}$,

- $C' = C_N \cup C_T$, where

$C_N = \{(ex'/cr') \mid \exists (ex/cr) \in C\colon$
$\quad lab(ex') = (lab(ex), \gamma_1 \ldots \gamma_{rank(ex)}, f) \in N_{info},$
$\quad f\colon [rank(ex)] \to [rank(ex')] \cup \{0\},$
$\quad \forall i \in [rank(ex)]\colon (ex[i] = \Diamond \wedge \gamma_i = X \wedge f(i) \neq 0 \wedge ex'[f(i)] = \Diamond) \vee$
$\quad\quad (ex[i] \in N \cup T \wedge \gamma_i = ex[i] \wedge (f(i) = 0 \vee ex'[f(i)] = ex[i])),$
$\quad lab(cr') = (lab(cr), \delta_1 \ldots \delta_{rank(cr)}, g) \in N_{info},$
$\quad g\colon [rank(cr)] \to [rank(cr')] \cup \{0\},$
$\quad \forall j \in [rank(cr)]\colon$
$\quad\quad (cr[j] \in V_H \wedge \delta_j = lab_H(cr[j]) \wedge (g(j) = 0 \vee cr'[g(j)] = cr[j])) \vee$
$\quad\quad (cr[j] \in \mathbb{N}_+ \wedge \delta_j = ex[cr[j]] \wedge (g(j) = 0 \vee cr'[g(j)] = f(cr[j]))),$
$\quad$ and (ex'/cr') is link-preserving $\}$

contains the connection instructions generating hyperedges with label in N_{info}, and

$C_T = \{(ex'/cr') \mid \exists (ex/cr) \in C\colon$
$\quad lab(ex') = (lab(ex), \gamma_1 \ldots \gamma_{rank(ex)}, f) \in N_{info},$
$\quad f\colon [rank(ex)] \to [rank(ex')] \cup \{0\},$
$\quad \forall i \in [rank(ex)]\colon (ex[i] = \Diamond \wedge \gamma_i = X \wedge f(i) \neq 0 \wedge ex'[f(i)] = \Diamond) \vee$
$\quad\quad (ex[i] \in N \cup T \wedge \gamma_i = ex[i] \wedge (f(i) = 0 \vee ex'[f(i)] = ex[i])),$
$\quad \forall j \in [rank(cr)]\colon (cr[j] \in V_H \wedge lab_H(cr[j]) \in T \wedge cr'[j] = cr[j]) \vee$
$\quad\quad (cr[j] \in \mathbb{N}_+ \wedge ex[cr[j]] \in T \wedge cr'[j] = f(cr[j])),$
$\quad$ and (ex'/cr') is link-preserving $\}$

the connection instructions which can generate hyperedges incident to terminally labelled nodes only.

By definition, NG' is link-preserving. As the nodes and node labels in NG' are exactly those of NG, there is a one-to-one correspondence between the derivations in NG and the derivations in NG'. More precisely, there is a derivation

$$\bullet\, S \;=\; H_0 \Rightarrow_{[v_1,p_1]} H_1 \Rightarrow_{[v_2,p_2]} \ldots H_{k-1} \Rightarrow_{[v_k,p_k]} H_k = H$$

in NG if and only if there is a derivation

$$\bullet\, S \;=\; H'_0 \Rightarrow_{[v_1,\lambda(p_1)]} H'_1 \Rightarrow_{[v_2,\lambda(p_2)]} \ldots H'_{k-1} \Rightarrow_{[v_k,\lambda(p_k)]} H'_k = H'$$

in NG', and H and H' have the same nodes with the same labels.

In the situation above, let $V \subseteq V_H = V_{H'}$ such that $v_i \in anc_{H_{i-1}}(V)$ for all $i \in [k]$, i.e., v_i is the ancestor in H_{i-1} of some node in V. For $lab_H(V) \cap N \neq \emptyset$, the definition of $\lambda(p)$ allows to show by induction on k:

(1) If $V \subseteq vset_H(e)$ for some hyperedge $e \in E_H$, then H' contains all sub-hyperedges e' of e with $vset_{H'}(e') = V$.

(2) If $V = vset_{H'}(e')$ for some hyperedge $e' \in E_{H'}$, then H contains a hyperedge e of which e' is a sub-hyperedge.

For $lab_H(V) \subseteq T$, claims (1) and (2) and the definition of $\lambda(p)$ imply:

(3) If V consists of the nodes incident to some hyperedge e, then $e \in E_H$ if and only if $e \in E_{H'}$.

Confluence of NG'. Suppose that NG' is not confluent. Then there is a derivation $\bullet\, S \Rightarrow^k_{P'} H'$ in NG' with k minimal such that there are a hyperedge e' in H' incident to two nonterminal nodes v_1, v_2, two productions $\lambda(p_i) = (lab_{H'}(v_i) ::= (H'_i, C'_i)) \in P'$ $(i = 1, 2)$, hypergraphs $H''_{ij} = H'[v_i/(H'_i, C'_i)][v_j/(H'_j, C'_j)]$ (where $i, j \in [2]$ and $i \neq j$) and a hyperedge $\bar{e}'$ generated from e' which belongs to, say, H'_{12}, but not to H'_{21}. As k is minimal with the properties above and NG' is link-preserving, only ancestors of nodes in $vset_{H'}(e')$ have been rewritten in the derivation $\bullet\, S \Rightarrow^k_{P'} H'$. Moreover, $\bar{e}'$ is generated from e' by two link-preserving connection instructions in C'_1 resp. C'_2, which implies that $anc_{H'}(vset_{H'_{12}}(\bar{e}')) = vset_{H'}(e')$.

Now consider the corresponding derivation $\bullet\, S \Rightarrow^k_{P} H$ in NG. By (2), there is at least one hyperedge e in H such that e' is one of the sub-hyperedges of e. Then by the definition of C'_1 and C'_2, there is an embedding hyperedge $\bar{e}$ in $H_{12} = H[v_1/(H_1, C_1)][v_2/(H_2, C_2)]$ generated from e such that $\bar{e}'$ is one of the sub-hyperedges of $\bar{e}$. By the confluence of NG, $\bar{e}$ also belongs to $H_{21} = H[v_2/(H_2, C_2)][v_1/(H_1, C_1)]$. Moreover, there is a hyperedge $\hat{e}$ in H which evolves into $\bar{e}$ in H_{21}. As

$$vset_{H'}(e') = anc_{H'}(vset_{H'_{12}}(\bar{e}')) \subseteq anc_H(vset_{H_{12}}(\bar{e})) = anc_H(vset_{H_{21}}(\bar{e})) \subseteq vset_H(\hat{e}),$$

claim (1) implies that H' contains a sub-hyperedge $\hat{e}'$ of $\hat{e}$ in H with $vset_{H'}(\hat{e}') = vset_{H'}(e')$. By construction of C_1' and C_2', this hyperedge $\hat{e}'$ evolves into the hyperedge $\bar{e}'$ in H_{21}', a contradiction. Therefore, NG' is confluent.

Equivalence of NG and NG'. Consider a derivation $\bullet s \Rightarrow_P^n H$ in NG with $lab_H(V_H) \subseteq T$. Let e be a hyperedge in E_H. As NG is confluent, the derivation steps can be reordered such that $\bullet s \Rightarrow_P^k G \Rightarrow_P^{n-k} H$ and k is minimal with $vset_H(e) \subseteq V_G$. The remote-freeness of NG implies $e \in E_G$ (Lemma 3.2), and by the minimality of k, in each of the k rewriting steps an ancestor of a node in $vset_G(e)$ was rewritten. Thus, for the corresponding derivation $\bullet s \Rightarrow_{P'}^k G' \Rightarrow_{P'}^{n-k} H'$ in NG' we have by claim (3) that $e \in E_G'$, which means $e \in E_{H'}$, too. As NG' is confluent, these derivation steps can be reordered to the derivation $\bullet s \rightarrow_{P'}^n H'$ corresponding to $\bullet s \Rightarrow_P^n H$, and still $e \in E_{H'}$. Hence, we have $E_H \subseteq E_{H'}$.

Similar reasoning yields that for each derivation $\bullet s \Rightarrow_{P'}^n H'$ in NG' with $lab_{H'}(V_{H'}) \subseteq T$ and the corresponding derivation $\bullet s \Rightarrow_P^n H$ in NG, $E_{H'} \subseteq E_H$.

Therefore for each $H \in \mathcal{H}_{N \cup T}$ with $lab_H(V_H) \subseteq T$, $H \in S(NG)$ if and only if $H \in S(NG')$, and in particular $H \in L(NG)$ if and only if $H \in L(NG')$. □

The second step towards the normal form of Theorem 3.3 consists of a way to get rid of sentential forms in which all nodes are terminal, but some hyperedges (of rank greater than k) are not.

3.6 Definition (non-blocking hNCE grammar)
Let $NG = (N, T, P, S)$ be an hNCE grammar and $H \in S(NG)$. A hyperedge $e \in E_H$ with $lab_H^*(att_H(e)) \in T^*$ and $lab_H(e) \in N$ is called a *blocking* hyperedge. The hypergraph H is called *terminal* if $lab_H(V_H) \subseteq T$. The grammar NG is *non-blocking* if no terminal sentential form of NG contains a blocking hyperedge.

The construction of a non-blocking normal form was first given by Skodinis and Wanke [SW95] for confluent eNCE grammars and reformulated for confluent edNCE grammars in [ER97]. The following theorem is a straightforward adaptation of Theorem 1.3.21 in [ER97].

3.7 Theorem (non-blocking normal form)
For every C-hNCE grammar NG an equivalent non-blocking C-hNCE grammar NG' can be constructed. If NG is remote-free or link-preserving, then so is NG'.

Proof. Let $NG = (N, T, P, S)$ be a C-hNCE grammar of rank k. Controlling the possible combinations of production applications allows to avoid the generation of blocking hyperedges. This can be done by adding to the label of a nonterminal node v some information (b, c), where $b \in \{true, false\}$ is *true* if and only if v cannot be derived into a terminal hypergraph with a blocking hyperedge, and c, consisting of simplified connection instructions, allows to select for the rewriting of v only productions with a specific type of connection relation.

In the construction of the normal-form grammar $NG' = (N', T, P', S')$, let, for a hypergraph with embedding $(H, C) \in \mathcal{HE}_{NUT}$ with H terminal, $info(H, C) = (b, c)$ denote the information with $b = true$ if and only if H contains no blocking hyperedge, and c the set of connection instructions (ex/cr) in C, but with each node $cr[i] \in V_H$ changed into its label $lab_H(cr[i])$. Moreover, let $S' = (S, (true, \emptyset))$ and $N' = (N \times Q) \uplus N$ with

$$Q = \{(b, c) \mid b \in \{true, false\}, c \subseteq (\Sigma \times \Sigma_\Diamond^*) \times (\Sigma \times ([k] \uplus T)^*),$$
$$\text{for all } (ex/cr) \in c: rank(ex) \in [k], rank(cr) \in [k] \cup \{0\}, \text{ and}$$
$$(i \in [rank(cr)] \wedge cr[i] \in [k] \Rightarrow cr[i] \in [rank(ex)] \wedge ex[cr[i]] \in \Sigma\}.$$

In order to construct the productions of NG', define for each production $p \in P$ the following function $\delta_p: Q^n \to Q$ (where n is the number of nonterminal nodes in the right-hand side hypergraph of p). Let $p = (X ::= (R, C))$, and let $v_1, \ldots, v_n$ be the nonterminal nodes of R, in that order. For all $i \in [n]$ and hypergraphs with embedding $(H_i, C_i) \in \mathcal{HE}_{NUT}$ with H_i terminal and $info(H_i, C_i) = (b_i, c_i)$, we want

$$\delta_p((b_1, c_1), \ldots, (b_n, c_n)) = info((R, C)[v_1/(H_1, C_1)] \ldots [v_n/(H_n, C_n)]).$$

For the computation of δ_p, it suffices to consider, for each $i \in [n]$, (an isomorphic copy of) the hypergraph with embedding (H_i, c_i) where H_i consists of T as node set, the identity as node labelling, one S-labelled hyperedge of rank 0 if $b_i = false$, and the empty hyperedge set otherwise.

Now P' contains, for every production $p \in P$ as above and $q_1, \ldots, q_n \in Q$, the production $prod(p, q_1, \ldots, q_n)$ with $(X, \delta_p(q_1, \ldots, q_n))$ as left-hand side and the right-hand side obtained from (R, C) by replacing the label X_i of v_i with (X_i, q_i) and each connection instruction (ex/cr) with all connection instructions (ex'/cr) where $ex'[i] \in \{ex[i]\} \times Q$ if $ex[i] \in N$ and $ex'[i] = ex[i]$ otherwise. It immediately follows that in every sentential form of NG', the nonterminal nodes have labels of the form $(X, (true, c))$, and there is no blocking hyperedge.

Clearly, every derivation $\bullet S' \Rightarrow_{P'}^n H'$ can be turned into a derivation $\bullet S \Rightarrow_P^n H$ simply by removing the additional information from the nonterminal node labels in the sentential forms and applied productions. In particular, this removal transforms H' into H. Consequently, NG' is confluent, and its generated language is a subset of $L(NG)$.

To show the converse inclusion, let, for a nonterminal node label X and a node x, $single(X, x)$ be the hypergraph with embedding (H, C) with $V_H = \{x\}$, $lab_H(x) = X$, $E_H = \emptyset$, and

$$C = \{(ex/cr) \mid (ex/cr) \text{ is form-preserving},$$
$$lab(ex) = lab(cr), \text{ and } rank(ex) = rank(cr) \leq k\}.$$

Then an induction on the length of the derivations allows to prove for a terminal hypergraph H that $single((X, q), x) \Rightarrow_{P'}^* (H, C')$ if (and only if) $single(X, x) \Rightarrow_P^*$

(H, C), $info(H, C) = q$, and C' contains all connection instructions (ex'/cr) with $(ex/cr) \in C$, $lab(ex') = lab(ex)$, $rank(ex') = rank(ex)$, and, for all $i \in [rank(ex)]$, $ex'[i] \in \{ex[i]\} \times Q$ if $ex[i] \in N$ and $ex'[i] = ex[i]$ otherwise. This holds in particular for $X = S$ and $q = (true, \emptyset)$, which implies $L(NG) \subseteq L(NG')$.

In the construction of NG', only node labels are changed. In particular, if a connection instruction of a production in NG' is remote or not link-preserving, such an instruction must occur in some production of NG, too. As a consequence, remote-freeness resp. link-preservation carries over from NG to NG'. □

Proof of Theorem 3.3. Let $NG = (N, T, P, S)$ be a C-hNCE$_{rf}$ grammar which generates a language of rank k. Construct the link-preserving normal form of NG as in the proof of Theorem 3.5, then the non-blocking normal form of the resulting grammar as in the proof of Theorem 3.7. The link-preserving and non-blocking C-hNCE grammar $NG' = (N', T, P', S')$ thus obtained is equivalent to NG.

As NG' is link-preserving, no hyperedge of rank greater than k can ever be transformed into an embedding hyperedge of rank at most k. In particular, it can never lead to a terminally labelled hyperedge of a hypergraph in $L(NG')$ because $rank(L(NG')) = k$, and neither can it lead to a nonterminally labelled blocking hyperedge because NG' is non-blocking. Therefore it is either implicitly deleted during some embedding, or it only occurs in a sentential form containing a nonterminal node which cannot be rewritten into a terminal hypergraph. Thus removing from the right-hand sides of the productions in P' all hyperedges of rank greater than k and all connection instructions whose creation parts specify such a hyperedge does not alter the language generated by the grammar. As NG' is link-preserving, all connection instructions whose creation part specifies a hyperedge of rank greater than k are thereby removed, too, so that the resulting grammar has rank k. □

Theorem 3.3 states in particular that every C-hNCE$_{rf}$ language of graphs can be generated by a grammar in which hyperedges of rank at most 2 are specified. Aiming at a C-edNCE normal form, we have to get rid of hyperedges of rank less than 2, such as hyperedges which are incident to exactly one nonterminal node.

3.8 Definition (pendant-free hNCE grammar)

Let $NG = (N, T, P, S)$ be an hNCE grammar. A hyperedge e in a hypergraph $H \in \mathcal{H}_{NUT}$ is a *pendant* if $vset_H(e) = \{v\}$ for some $v \in V_H$ with $lab_H(v) \in N$. A production $X ::= (H, C)$ in P is *pendant-free* if H does not contain a pendant and C does not contain a connection instruction (ex/cr) of one of the following three forms:

$ex[i] = \Diamond$ for all $i \in [rank(ex)]$,
there is $v \in V_H$ with $cr[i] = v$ for all $i \in [rank(cr)]$ and $lab_H(v) \in N$, or
there is $j \in [rank(ex)]$ with $cr[i] = j$ for all $i \in [rank(cr)]$ and $ex[j] \in N$.

The grammar NG is pendant-free if all its productions are.

3.9 Theorem (pendant-free normal form)

For every link-preserving hNCE grammar NG an equivalent link-preserving hNCE grammar NG' can be constructed which is pendant-free. If NG is confluent or non-blocking, then so is NG'.

Proof. Let $NG = (N, T, P, S)$ be a link-preserving hNCE grammar. As $\Diamond$ occurs at most once in the existence part of a link-preserving connection instruction, every pendant of rank greater than 1 will be deleted when a link-preserving production is applied to its incident node. We may therefore assume for the right-hand sides of all productions in P that the rank of the pendants in the hypergraph is 1, and all connection instructions which may create a pendant are of the form $(\alpha, \Diamond/\beta, v)$ with $v \in V_H$ and $lab_H(v) \in N$.

Now construct an hNCE grammar $NG' = (N', T, P', S')$ such that the information on the pendants of each node is stored in the (nonterminal) label of the node and the pendants are discarded. More precisely, let $Q = \mathcal{P}(N \cup T)$. An element $q = \{\beta_1, \ldots, \beta_l\}$ of Q describes l pendants of rank 1, where the ith pendant is labelled β_i. For each q as above, let $hg(q, x, X)$ be the hypergraph consisting of an X-labelled node x with pendants as described by q, i.e. $hg(q, x, X) = (V, E, lab)$ with $V = \{x\}$ and $E = \{(\beta_i, x) \mid i \in [l]\}$.

Let $p = (X ::= (R, C))$ be a production of NG, and let $v_1, \ldots, v_n$ be the non-terminal nodes of R, in that order, with labels $X_1, \ldots, X_n$, respectively. Construct a function $\delta_p \colon Q \to Q^n$ such that for $q \in Q$ and $\delta_p(q) = (q_1, \ldots, q_n)$, q_i describes the pendants of the node v_i in the hypergraph $R_q = hg(q, x, X)[x/(R, C)]$.

Let $N' = (N \times Q) \uplus N$. The productions of NG' are defined as follows. For each production p of NG as above and every $q \in Q$, P' contains the production $prod(p, q) = ((X, q) ::= (R', C'))$, where R' is the hypergraph obtained from R_q by discarding all pendants and replacing, for all $i \in [n]$, the label X_i of v_i by (X_i, q_i), and C' is obtained from C by discarding all connection instructions (ex/cr) with $rank(ex) = 1$, and replacing each connection instruction (ex/cr) which is left with all connection instructions (ex'/cr) obtained from (ex/cr) by substituting, for each $i \in [rank(ex)]$, some $ex'[i] \in \{ex[i]\} \times Q$ for $ex[i] \in N$. As p is link-preserving, this implies that $prod(p, q)$ is link-preserving and pendant-free. Finally, let $S' = (S, \emptyset)$ be the start symbol of NG'.

There is a one-to-one correspondence between a derivation $\bullet S \Rightarrow_P^* H$ in NG and a derivation $\bullet (S, \emptyset) \Rightarrow_{P'}^* H'$ in NG', and H' differs from H in that it is pendant-free and a nonterminal node v' contains in its label the information on the pendants of its corresponding node in H. Consequently, $L(NG) = L(NG')$, NG' is confluent if NG is, and NG' generates a sentential form with a blocking hyperedge only if NG does. $\square$

Now we are ready to show that every graph-generating C-hNCE$_{rf}$ grammar has a C-edNCE normal form.

3.10 Theorem (C-edNCE normal form)

For every C-hNCE$_{rf}$ grammar NG with $L(NG) \in \mathcal{LG}$, an equivalent C-edNCE grammar can be constructed.

Proof. Let $NG = (N, T, P, S)$ be a C-hNCE$_{rf}$ grammar with $L(NG) \in \mathcal{LG}$. To construct an equivalent C-edNCE grammar $NG' = (N', T, P', S')$, transform NG into its link-preserving normal form NG_1 (Theorem 3.5), NG_1 into its non-blocking normal form NG_2 (Theorem 3.7), and NG_2 into its pendant-free normal form NG_3 (Theorem 3.9). An immediate consequence of the normal-form constructions is that $NG_3 = (N_3, T, P_3, S_3)$ is confluent, link-preserving, non-blocking, and pendant-free.

Consider a graph $G = (V_G, E_G, lab_G)$ in $L(NG_3)$. As NG_3 is pendant-free and link-preserving, each edge $e \in E_G$ is either (a copy of) an edge in the right-hand side of a production, or transformed from another edge by a (succession of) link-preserving connection instruction(s) where the existence and creation parts have rank 2. Thus, G can still be generated if in the right-hand sides of the productions of NG_3 all hyperedges of rank other than 2 and all connection instructions of a different form are deleted.

Formally, define NG' such that $N' = N_3$, $S' = S_3$, and P' contains, for each production $(X ::= (H, C)) \in P_3$ with $H = (V_H, E_H, lab_H)$, a production $X ::= (H', C')$ with $H' = (V_{H'}, E_{H'}, lab_{H'})$ such that $V_{H'} = V_H$, $lab_{H'}(v) = lab_H(v)$ for all $v \in V_{H'}$, $E_{H'} = \{e \in E_H \mid att_H(e) = v_1 v_2 \text{ with } v_1 \neq v_2\}$, and $C' = \{(ex/cr) \in C \mid rank(ex) = 2 = rank(cr)\}$.

Then NG' is a confluent edNCE grammar, and $L(NG_3) \subseteq L(NG')$ by the reasoning above. Moreover, if there was a derivation $\bullet S' \Rightarrow_{P'}^* G$ with $G \in L(NG')$ but $G \notin L(NG_3)$, then the corresponding derivation $\bullet S_3 \Rightarrow_{P_3}^* G'$ must have produced a blocking hyperedge in G', which cannot be as NG_3 is non-blocking. So, $L(NG') = L(NG_3) = L(NG)$. $\qquad\square$

3.11 Corollary

$\mathcal{L}(\text{C-edNCE}) = \mathcal{L}(\text{C-hNCE}_{rf}) \cap \mathcal{LG}$.

Proof. By Definition 2.10 and Lemma 2.18, every C-edNCE grammar is a remote-free C-hNCE grammar generating a graph language, which means $\mathcal{L}(\text{C-edNCE}) \subseteq \mathcal{L}(\text{C-hNCE}_{rf}) \cap \mathcal{LG}$. The converse holds by Theorem 3.10. $\qquad\square$

3.2 Simulation of Hyperedge Rewriting

Hyperedge rewriting is the earliest context-free hypergraph rewriting technique. It was introduced independently by Bauderon and Courcelle [BC87] and Habel and Kreowski [HK87], thoroughly studied in [Hab92a], and a recent survey is [DHK97]. The idea of this approach is to take hyperedges as nonterminal items. Such a

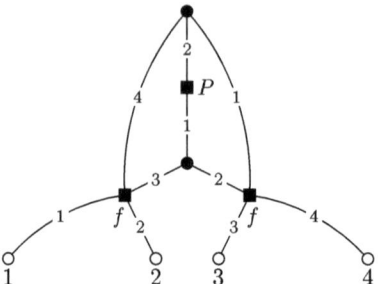

Figure 3.2: The hypergraph with external nodes (M_2, ext_2)

hyperedge is a placeholder which specifies a sequence of nodes in the hypergraph it belongs to. The same sequence of nodes may be specified by more than one nonterminal item, so this rewriting technique is defined on the basis of multiple hypergraphs. A hyperedge can be rewritten by putting a hypergraph in its place, which is subsequently linked to the host hypergraph by gluing some of its nodes (called external nodes) with the attachment nodes of the rewritten hyperedge. Thus, hyperedge rewriting is an example of the gluing approach to (hyper)graph rewriting. Moreover, it is context-free in the sense of [Cou87].

Engelfriet and Rozenberg showed in [ER90] that HR grammars and a subtype of edNCE grammars which are called boundary edNCE grammars of bounded non-terminal degree (B-edNCE$_{bntd}$ grammars) have the same (hyper)graph generating power if hypergraphs are seen as bipartite graphs. By contrast, the idea pursued in this section is to simulate hyperedge replacement via node replacement while maintaining the complex structure of hyperedges. Not surprisingly, the constructed hNCE grammar is also boundary and therefore confluent (see Section 2.4). Unlike [Kle96], only HR grammars which generate simple hypergraphs are considered here; in return, this allows to construct hNCE grammars which generate the same languages (without encoding of multiple hyperedges).

In this section, every label $\alpha \in \Sigma$ is supposed to have a type $type(\alpha) \in \mathbb{N}$. For a multiple hypergraph $M = (V_M, E_M, lab_M, att_M)$, nodes are unlabelled (i.e. $lab_M(V_M) = \{*\}$) and the rank of a hyperedge is reflected in the type of its label (i.e. $type(lab_M(e)) = rank_M(e)$ for all $e \in E_M$).

3.12 Definition (hypergraph with external nodes)
A *hypergraph with external nodes* over Σ is a pair (M, ext) where M is a multiple hypergraph over Σ and $ext \in V_M{}^*$ is a sequence of pairwise distinct *external* nodes.[1]

---

[1]For hyperedge rewriting, it is not strictly necessary to require that the nodes in the sequence *ext* be pairwise distinct, but such a *repetition-free* normal form can always be constructed, see [Hab92a, Theorem 4.6 in Chapter I].

The set of external nodes is denoted by EXT. The set of all hypergraphs with external nodes over Σ is denoted by $\mathcal{MX}_\Sigma$.

Two hypergraphs with external nodes $(M, ext), (M', ext') \in \mathcal{MX}_\Sigma$ are isomorphic if there is an isomorphism $f : M \to M'$ such that $f^*(ext) = ext'$. A hypergraph with external nodes (M, λ) is considered to be the same as the multiple hypergraph M; thus $\mathcal{M}_\Sigma \subseteq \mathcal{MX}_\Sigma$.

3.13 Example (hyperedge rewriting)
Figure 3.2 shows a hypergraph with external nodes (M_2, ext_2), where the ith node in ext_2 has i written next to it $(i \in [4])$; the external nodes are white.

Figure 3.3 illustrates the substitution of (M_2, ext_2) for the upper f-labelled hyperedge e of the multiple hypergraph M_1:

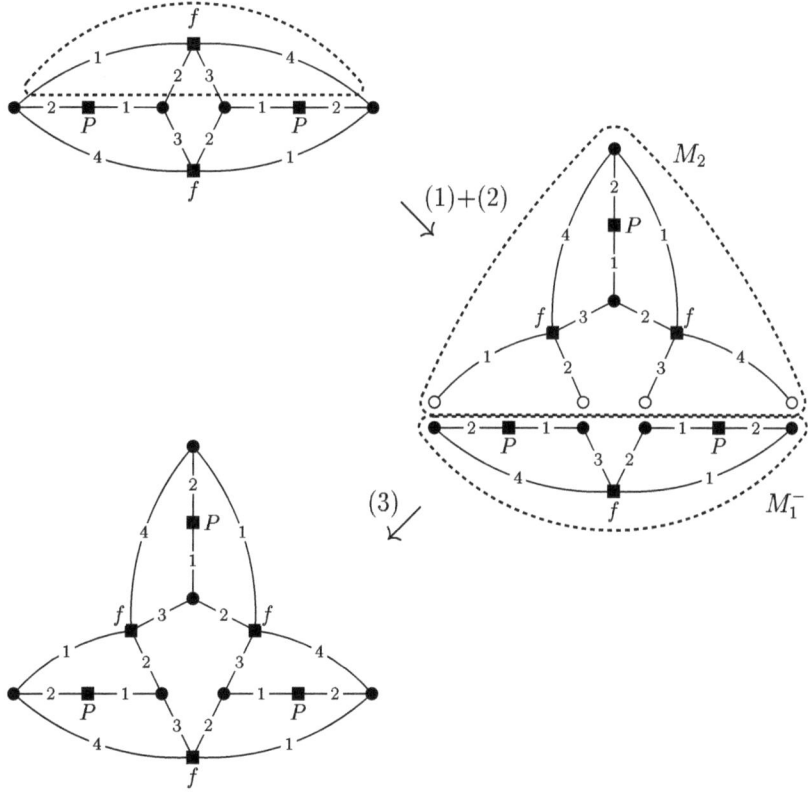

Figure 3.3: Substituting (M_2, ext_2) for a hyperedge

(1) REMOVE e (but without its attachment nodes), yielding the remainder M_1^- of M_1;

(2) ADD M_2 to M_1^-; and

(3) CONNECT M_2 and M_1^- by fusing the ith external node of M_2 with the ith attachment node of e. ∎

Formally, hyperedge rewriting is defined as follows.

3.14 Definition (HR rewriting)

Let (M_1, ext_1), (M_2, ext_2) be two disjoint hypergraphs with external nodes and e a hyperedge in M_1 with $rank_1(e) = |ext_2|$. Then $(M_1, ext_1)[e/(M_2, ext_2)]$ is the hypergraph with external nodes (M_3, ext_3) defined as follows, where $M_i = (V_i, E_i, lab_i, att_i)$ for $i \in [3]$:

- $V_3 = V_1 \cup (V_2 \smallsetminus EXT_2)$,

- $E_3 = (E_1 \smallsetminus \{e\}) \cup E_2$,

- $lab_3(e') = \begin{cases} lab_1(e') & \text{if } e' \in E_1 \smallsetminus \{e\}, \\ lab_2(e') & \text{if } e' \in E_2, \end{cases}$

- $rank_3(e') = \begin{cases} rank_1(e') & \text{if } e' \in E_1 \smallsetminus \{e\}, \\ rank_2(e') & \text{if } e' \in E_2, \end{cases}$

- $att_3(e', i) = \begin{cases} att_1(e', i) & \text{if } e' \in E_1 \smallsetminus \{e\}, \\ att_1(e, j) & \text{if } e' \in E_2 \text{ and } att_2(e', i) = ext_2(j), \\ att_2(e', i) & \text{if } e' \in E_2 \text{ and } att_2(e', i) \in V_2 \smallsetminus EXT_2, \end{cases}$

  for $i \in [rank_3(e')]$,

- $ext_3 = ext_1$.

A hyperedge-rewriting hypergraph grammar (HR grammar for short) is a tuple $HG = (N, T, P, S)$ where N and T are finite, disjoint sets of nonterminal and terminal symbols respectively, $P \subseteq N \times \mathcal{MX}_{N \cup T}$ is a finite set of productions, and $S \in N$ is the initial nonterminal. The axiom of HG is the multiple hypergraph ∎S consisting of one S-labelled hyperedge of rank 0 and no nodes. Derivations, sentential forms, and generated hypergraphs are defined as usual. An HR grammar HG is *hypergraph-generating* if the hypergraphs in the generated language are simple, i.e. $L(HG) \in \mathcal{LH}$. The class of hypergraph languages which can be generated with HR grammars is denoted by $\mathcal{L}(\mathrm{HR})$.

3.15 Example (HR grammar)

The set DOTTEDTREES of all binary trees with one additional node which is adjacent to all other nodes can be generated by the HR grammar $HG = (\{S, X\}, \{b, d, *\}, \{p_1, p_2, p_3\}, S)$ with the productions as follows:

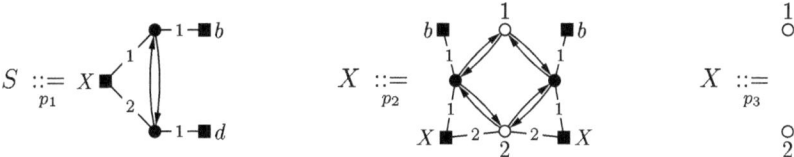

The distinguished node is the one incident to a d-labelled hyperedge, the first incident node of each X-labelled hyperedge is a leaf of the tree to which two children may be added with p_2, and the second incident node is the distinguished one.

A sample derivation of HG is shown in Figure 3.4, where each couple of edges ●⟶● is represented by a straight line ●——●, and the tentacle numbers of hyperedges of rank one are omitted. Moreover, the tree is indicated by bold edges.

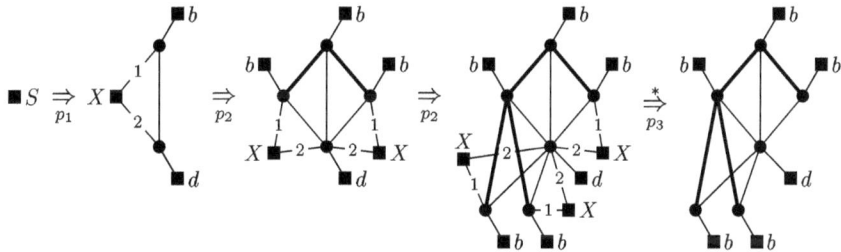

Figure 3.4: A derivation in grammar HG yielding the dotted tree with 5+1 nodes

The grammar HG is the grammar G_4 of [CER93]. In the same reference, a variant of this grammar is considered, obtained by adding an $n + 1$st 'private' incident node to each of the terminal hyperedges in the right-hand sides of the productions. The language generated by this grammar is referred to by $plus(\text{DOTTEDTREES})$.   ∎

Clearly, HR grammars (like hNCE grammars) do not generate hypergraph languages of unbounded rank.

3.16 Fact
For every HR grammar HG, $rank(L(HG))$ is bounded.

Moreover, we need only consider HR grammars which are *reduced*, i.e. every nonterminal symbol occurs in some sentential form, and each sentential form can be derived into a terminal hypergraph. The reduced normal form of an HR grammar can be constructed just as in the case of context-free string grammars [Hab92a, Chapter IV, Theorem 1.2] (cf. [HU79]).

3.17 Fact
For every HR grammar, an equivalent reduced HR grammar can be constructed.

In order to compare the generative power of hyperedge versus node rewriting, we consider in the sequel hypergraph-generating HR grammars.[2] These grammars have the following property.

3.18 Lemma
Let HG be a reduced hypergraph-generating HR grammar. Then no sentential form of HG contains two parallel terminal hyperedges.

Proof. Let $HG = (N, T, P, S)$ be a reduced HR grammar. Then for all derivations ∎ $S \Rightarrow_P^* M$, there is a derivation $M \Rightarrow_P^* H$ with $H \in \mathcal{M}_T$. If M contains parallel hyperedges with terminal label, then these hyperedges also belong to H as they cannot be deleted by hyperedge replacement. Thus H is not simple, a contradiction to HG being hypergraph-generating.   □

The main result of this section is as follows.

3.19 Theorem (simulation of HR rewriting)
For every hypergraph-generating HR grammar, an equivalent C-hNCE$_{rf}$ grammar can be constructed.

Outline of the proof. The idea is to add a node to each nonterminal hyperedge which will be rewritten instead of the hyperedge. In addition, the external nodes in the right-hand side of an HR production are removed, and their incident hyperedges generated as embedding hyperedges.   □

In order to prove that there is an equivalent C-hNCE grammar which is indeed remote-free, we first need a particular normal form of hypergraph-generating HR grammars.

Let $HG = (N, T, P, S)$ be an HR grammar and (M, ext) a hypergraph with external nodes over $N \cup T$. A hyperedge $e \in E_M$ with $lab_M(e) \in T$ and $vset_M(e) \subseteq EXT$ is called an *external edge*. A hypergraph with external nodes is *external edge-free* if it has no external edge, and an HR grammar is external edge-free if all right-hand sides of its productions are.

Every hypergraph-generating HR grammar can be transformed bottom-up into an external edge-free HR grammar: the basic idea is to delete all external edges in the right-hand side of a production, storing in its left-hand side the information that

---

[2]Alternatively, hNCE rewriting could be extended to multiple hypergraphs for a common basis to compare the mechanisms.

these hyperedges must be generated prior to applying the production (see [CER93, Lemma 7.7] where external edges are called port parasites, [Vog93, Proposition 4.3], and also [Eng97, Theorem 3.17]).

3.20 Fact
For every hypergraph-generating HR grammar, an equivalent external edge-free HR grammar can be constructed.

Proof of Theorem 3.19. Let $HG = (N, T, P, S)$ be an HR grammar with $L(HG) \in \mathcal{LH}$. We construct an hNCE grammar $NG = (N, T, P', S)$ where essentially each nonterminal hyperedge in a hypergraph of HG gets an additional attachment node which takes the label from the hyperedge. Moreover, each production of NG corresponds to a production of HG where the external nodes of the right-hand side have been deleted and hyperedges which were incident to such a node are generated via the connection relation. Figure 3.5 shows an example for this transformation.

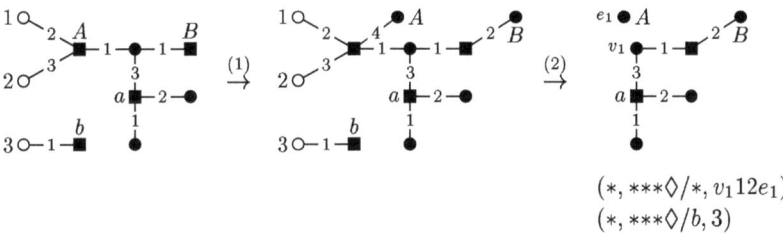

$$(*, ***\lozenge/*, v_1 12 e_1)$$
$$(*, ***\lozenge/b, 3)$$

Figure 3.5: Hypergraph transformation: (1) Adding a node to each nonterminal hyperedge (2) Replacing external nodes with connection instructions

Let (M, ext) be a hypergraph with external nodes, where $M = (V_M, E_M, lab_M, att_M)$ is a multiple hypergraph over $N \cup T$. Then (M, ext) is transformed into the hypergraph with embedding $(H, C) = \tau(M, ext)$ with $H = (V_H, E_H, lab_H)$ as follows:

- $V_H = (V_M \smallsetminus EXT) \cup \{\bar{e} \mid e \in E_M \text{ and } lab_M(e) \in N\}$,

- $E_H = \{(lab_M(e), att_M(e)) \mid \exists e \in E_M : vset_M(e) \cap EXT = \emptyset, lab_M(e) \in T\}$
$\cup \{(*, att_M(e) \cdot \bar{e}) \mid \exists e \in E_M : vset_M(e) \cap EXT = \emptyset, lab_M(e) \in N\}$,

- $lab_H(v) = \begin{cases} lab_M(e) & \text{if } v = \bar{e} \text{ and } e \in E_M, \\ * & \text{otherwise,} \end{cases}$ for all $v \in V_H$,

- $C = \{(ex/cr) \mid lab(ex) = *, \; rank(ex) = |ext| + 1,$
  $ex[i] = *$ for all $i \in [|ext|], \; ex[|ext| + 1] = \Diamond,$
  $\exists e \in V_M : \; vset_M(e) \cap EXT \neq \emptyset,$
  $(lab_M(e) \in T \wedge lab(cr) = lab_M(e) \wedge rank(cr) = rank_M(e))$ or
  $(lab_M(e) \in N \wedge lab(cr) = * \wedge rank(cr) = rank_M(e) + 1 \wedge$
  $\qquad\qquad\qquad\qquad\qquad\qquad\qquad cr[rank(cr)] = e),$

$$cr[i] = \begin{cases} j & \text{if } att_M(e, i) = ext(j), \\ att_M(e, i) & \text{otherwise} \end{cases}$$

$$\text{for all } i \in [rank_M(e)] \; \}.$$

By Fact 3.17 and Lemma 3.18, the multiple hypergraphs and sentential forms of HG are exclusively multiple hypergraphs without parallel terminal hyperedges. For this kind of multiple hypergraphs, the transformation τ is an encoding. Moreover, as by Lemma 3.20 the right-hand side (R, ext) of each production in HG may be assumed to be external edge-free, there is no remote connection instruction in $\tau(R, ext)$. (Compare the example of Figure 3.5: the b-labelled hyperedge is an external edge and responsible for the only connection instruction $(*, ***\Diamond/b, 3)$ in the encoded hypergraph which is remote.)

Let $P' = \{X ::= \tau(R, ext) \mid (X ::= (R, ext)) \in P\}$. By induction on the length of the derivations, we have for $H \in \mathcal{H}_T$:

$$(M, ext) \Rightarrow^*_P (H, ext) \quad \text{if and only if} \quad \tau(M, ext) \Rightarrow^*_{P'} \tau(H, ext),$$

which implies for all $H \in \mathcal{H}_T$:

$$\blacksquare S = (\blacksquare S, \lambda) \Rightarrow^*_P (H, \lambda) = H \quad \text{if and only if} \quad \tau(\blacksquare S, \lambda) \Rightarrow^*_{P'} \tau(H, \lambda) = H.$$

As $rank(S) = 0$, we have for all productions $(S ::= (R, ext)) \in P$: $ext = \lambda$ and thus $\tau(R, ext) = (R', \emptyset)$, i.e. the right-hand side of a production in P' with S as left-hand side has an empty connection relation. Thus for all hypergraphs $H \in \mathcal{H}_T$:

$$\tau(\blacksquare S) = \blacksquare{-}1{-}\bullet S \Rightarrow^*_{P'} H \quad \text{if and only if} \quad \bullet S \Rightarrow^*_{P'} H,$$

which concludes the proof that $L(HG) = L(NG)$.

Each nonterminal node in a hypergraph of sentential form of NG has only $*$-labelled, i.e. terminal, neighbours. Thus, NG is boundary, and in particular confluent by Lemma 2.21. $\qquad\square$

3.21 Corollary
$\mathcal{L}(HR) \cap \mathcal{LH} \subsetneq \mathcal{L}(C\text{-hNCE}_{rf}).$

Proof. By Theorem 3.19, $\mathcal{L}(HR) \cap \mathcal{LH} \subseteq \mathcal{L}(C\text{-hNCE}_{rf})$. Example 2.9.1 shows that COMPLETE $\in \mathcal{L}(C\text{-hNCE}_{rf})$, and obviously COMPLETE $\in \mathcal{LH}$, but COMPLETE $\notin \mathcal{L}(HR)$ by Theorem 2.6 in [Hab92a, Chapter IV]. $\qquad\square$

3.3   Simulation of Separated Handle Rewriting

Handle-rewriting hypergraph grammars were developed by Courcelle, Engelfriet and Rozenberg as a generalisation of hyperedge rewriting [CER93]. The idea is to replace a *handle*, i.e. a hyperedge e together with its incident nodes, with a hypergraph that has so-called port nodes, where a port node is a node which carries one or more natural numbers. The i-ports (for some $i \in \mathbb{N}_+$) play the role of the ith external node in hyperedge rewriting: for each i-port v, a copy of each hyperedge incident to the ith attachment node of e is created and made incident to v instead of the i-port. Thus handle rewriting allows to multiply (or delete) hyperedges incident to the nonterminal item, but unlike hNCE rewriting it does not allow dynamic relabelling or a change in the rank resp. attachment sequence of these hyperedges.

3.22 Definition (hypergraph with ports)
Let Σ be a set of labels. A *hypergraph with ports* over Σ is a pair $(H, port)$ where $H = (V_H, E_H, lab_H)$ is a hypergraph over Σ with $lab_H(V_H) = \{*\}$ (i.e. the nodes of H are unlabelled) and $rank_H(e) \geq 1$ for all $e \in E_H$, and $port$ is a finite subset of $\mathbb{N}_+ \times V_H$. As the nodes in a hypergraph with ports are unlabelled, we will drop the component lab_H for these hypergraphs and write $H = (V_H, E_H)$. For $(i, v) \in port$, v is called an i-port and i a port number of v. Moreover, $port(i)$ denotes the set $\{v \in V_H \mid (i, v) \in port\}$, and $port^{-1}(v)$ denotes the set $\{i \in \mathbb{N}_+ \mid (i, v) \in port\}$. The set of all hypergraphs with ports over Σ is denoted by $\mathcal{HP}_\Sigma$.

Two hypergraphs with ports $(H, port), (H', port') \in \mathcal{HP}_\Sigma$ are isomorphic if there is an isomorphism $f \colon H \to H'$ such that $port' = \{(i, f(v)) \mid (i, v) \in port\}$. A hypergraph with ports $(H, \emptyset)$ is considered to be the same as the hypergraph H; in this sense we have $\mathcal{H}_\Sigma \subseteq \mathcal{HP}_\Sigma$.

3.23 Example (handle rewriting)
Figure 3.6 shows a hypergraph with ports $(H_2, port_2)$, where an i-port has i written next to it $(i = 2, 3)$.

Figure 3.7 illustrates the substitution of $(H_2, port_2)$ for the rightmost t-labelled hyperedge e of the hypergraph H_1:

(1) REMOVE e together with its attachment nodes and all their incident hyperedges, yielding the remainder H_1^- of H_1;

(2) ADD H_2 to H_1^-; and

(3) CONNECT H_2 and H_1^- according to the relation $port_2$, by creating for each node v of H_2 with $v \in port(i)$ a copy of each hyperedge $e' \in E_1$ with $e' \neq e$ and $att_1(e, i) \in vset_1(e')$ and making this copy incident to v instead of $att_1(e, i)$.

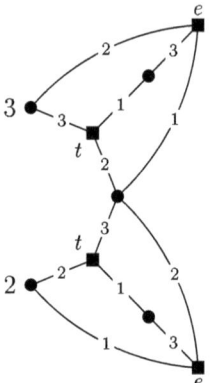

Figure 3.6: The hypergraph with ports $(H_2, port_2)$

For example, copies of the lower e- and t-labelled hyperedges on the left of H_1 are created to be incident with the 2-port of $(H_2, port_2)$. The e-labelled hyperedge on the right of H_1 is not recreated because it is incident to $att_1(e, 1)$ and there is no 1-port in $(H_2, port_2)$. ∎

Formally, handle rewriting is defined as follows.

3.24 Definition (HH rewriting)
Let $(H_1, port_1)$, $(H_2, port_2)$ be two disjoint hypergraphs with ports and e a hyperedge in H_1. Then $(H_1, port_1)[e/(H_2, port_2)]$ is the hypergraph with ports $(H_3, port_3)$ defined as follows, where $H_i = (V_i, E_i)$ for $i \in [3]$:

- $V_3 = (V_1 \smallsetminus vset_1(e)) \cup V_2$,

- $E_3 = \quad \{(\alpha, v_1 \ldots v_n) \mid (\alpha, u_1 \ldots u_n) \in E_1 \smallsetminus \{e\},$
  $\qquad\qquad$ for all $i \in [n]$: either $u_i \notin vset_1(e)$ and $v_i = u_i$, or
  $\qquad\qquad \exists j \in [rank_1(e)]$: $u_i = att_1(e, j)$ and $v_i \in port_2(j)\}$
  $\qquad \cup E_2,$

- $port_3 = \{(i, v) \mid \exists u \in V_1 : (i, u) \in port_1$ and
  $\qquad\qquad$ either $u \notin vset_1(e)$ and $v = u$, or
  $\qquad\qquad \exists j \in [rank_1(e)]$: $u = att_1(e, j)$ and $v \in port_2(j)\}$.

A handle-rewriting hypergraph grammar (HH grammar for short) is a tuple $HG = (N, T, P, S)$ where N and T are finite, disjoint sets of nonterminal and terminal symbols respectively, $P \subseteq N \times \mathcal{HP}_{NUT}$ is a finite set of productions, and $S \in N$ is the initial nonterminal. The axiom of HG is the hypergraph ●–1–■ S consisting of one S-labelled hyperedge of rank 1 and one node. Derivations, sentential forms, and generated hypergraphs are defined as usual.

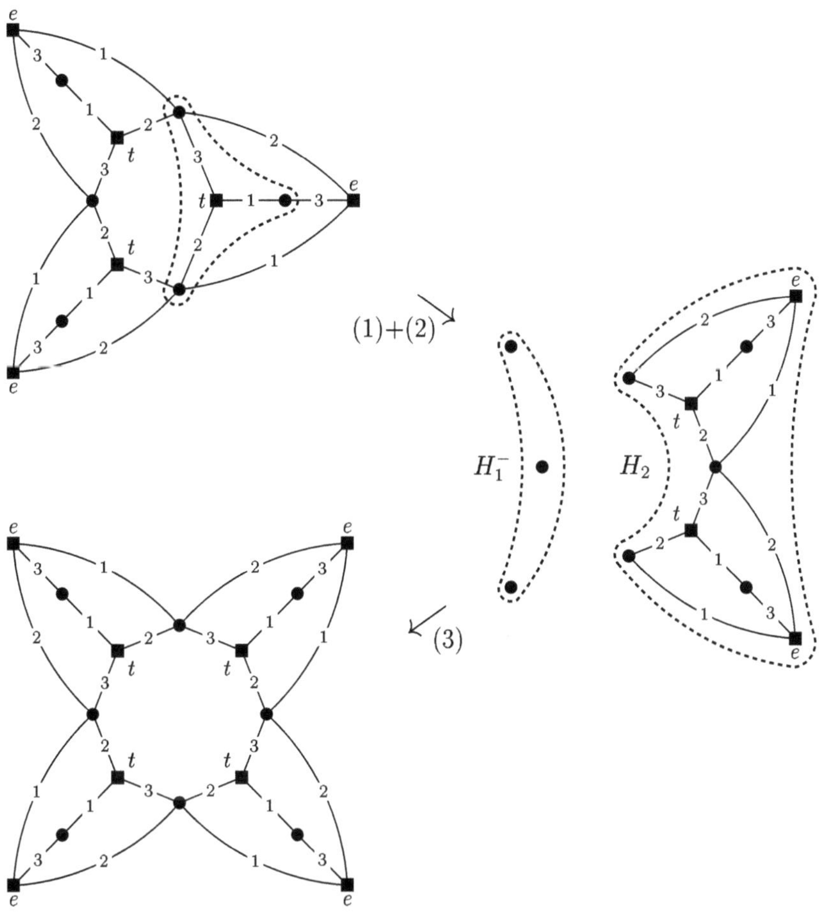

Figure 3.7: Substituting $(H_2, port_2)$ for a hyperhandle

3.25 Example (HH grammar)

The set ADDCOMPLETE of all complete graphs with a private hyperedge of rank one made incident to each of the nodes can be generated by the S-HH grammar $HG = (\{S\}, \{a, *\}, \{p_1, p_2\}, S)$ with the productions as follows:

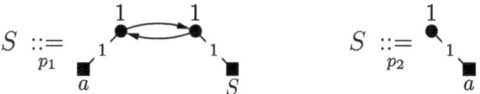

A sample derivation of HG is shown in Figure 3.8.   ∎

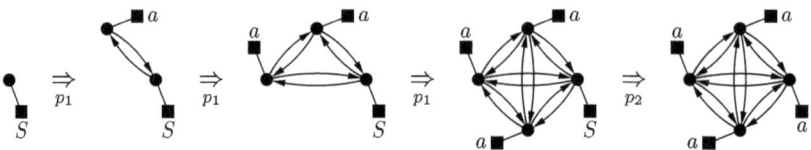

Figure 3.8: A derivation in grammar HG yielding the 'complete graph' with four nodes

HH grammars do not generate languages of unbounded rank.

3.26 Fact
For every HH grammar HG, $rank(L(HG))$ is bounded.

If two nonterminally labelled hyperedges have a common incident node, then rewriting one of the handles automatically affects the other. Excluding these situations leads to the concept of separated handle rewriting.

3.27 Definition (separated HH grammar)
Let $HG = (N, T, P, S)$ be a HH grammar. A hypergraph with ports $(H, port)$ over $N \cup T$ is *separated* if $vset_H(e) \cap vset_H(e') \neq \emptyset$ implies $e = e'$ for all nonterminal hyperedges $e, e' \in E_H$. The grammar HG is separated (an S-HH grammar for short) if all the right-hand sides of its productions are. The class of hypergraph languages which can be generated with S-HH grammars is denoted by $\mathcal{L}(\text{S-HH})$.

Separated handle rewriting is, just as hyperedge rewriting, a naturally context-free approach.

3.28 Fact
Let HG be an S-HH grammar. Then all sentential forms of HG are separated. This implies that HG is context-free.

The following result is a straightforward generalisation of Lemma 4.1 in [CER93], where an arbitrary graph-generating S-HH grammar is transformed into an equivalent C-edNCE grammar.

3.29 Theorem (simulation of S-HH rewriting)
For every S-HH grammar, an equivalent C-hNCE$_{rf}$ grammar can be constructed.

Proof. Let $HG = (N, T, P, S)$ be an S-HH grammar amd m the maximal rank of a hyperedge occurring in HG. Construct an hNCE grammar $NG = (N', T, P', S)$ using the following idea: Each nonterminal handle is contracted into a nonterminal node carrying the label of the handle, and each hyperedge incident to a nonterminal handle gets a new label into which the information on this incidence is encoded. An example is shown on Figure 3.9, where X, Y are nonterminal and α, β terminal

labels. The X- and Y-labelled handles of the hypergraph on the left are contracted
into the big nodes of the hypergraph on the right, the α-labelled hyperedge which is
not incident to a nonterminal label remains unchanged, and the β-labelled hyperedge e gets a new label augmented by a sequence of length $rank(e)$ where the entry
j at position i means that $att(e, i)$ was the jth attachment node of a nonterminal
hyperedge, and $j = 0$ if $att(e, i)$ is not incident to a nonterminal hyperedge.

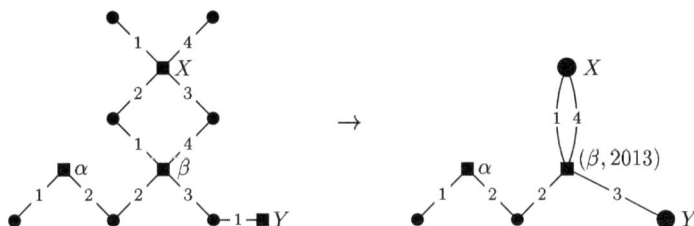

Figure 3.9: Translating a hypergraph in $\mathcal{HP}_{NUT}$ into a hypergraph in $\mathcal{HE}_{N'UT}$

As we consider only separated hypergraphs, two distinct handles are never
contracted into the same node, and each attachment node of a terminal hyperedge
belongs to at most one nonterminal handle. Moreover, a new nonterminal node
knows by the label of an incident hyperedge just how this hyperedge was connected
to the original handle, so that connection instructions can be used to generate the
same connecting hyperedges as the ports would have done.

For the formal construction of NG, let $M = T \times \bigcup_{i \in [m]}([m] \cup \{0\})^i$, $M_T = T \times \bigcup_{i \in [m]} \{0\}^i$, and $M_N = M \smallsetminus M_T$. The new nonterminal alphabet is $N' = N \cup M_N$,
where N contains the node labels and M_N the hyperedge labels. For notational
simplicity, any element $(\alpha, 0 \ldots 0) \in M_T$ will be identified with α, so that every
hyperedge will be labelled with some element $(\alpha, i_1 \ldots i_k) \in M$, where $i_j > 0$
if the jth incident node was the i_jth node of a nonterminal handle and $i_j = 0$
otherwise. Moreover, for each terminal node v in a hypergraph $H \in \mathcal{H}_{NUT}$, we let
$att_H(v, 0) = v$.

Formally, the translation $\tau \colon \mathcal{HP}_{NUT} \to \mathcal{HE}_{N'UT}$ is defined as follows. For a
separated hypergraph with ports $(H, port) \in \mathcal{HP}_{NUT}$ with $H = (V_H, E_H)$ and
hyperedge ranks at most m, $\tau(H, port)$ is the hypergraph with embedding $(H', C) \in \mathcal{HE}_{N'UT}$, where $H' = (V_{H'}, E_{H'}, lab_{H'})$ and

- $V_{H'} = \{e \in E_H \mid lab_H(e) \in N\} \cup \{v \in V_H \mid lab_H(eset_H(v)) \subseteq T\}$,

- $E_{H'} = \{((\alpha, i_1 \ldots i_k), v_1 \ldots v_k) \mid$
  $(\alpha, att_H(v_1, i_1) \ldots att_H(v_k, i_k)) \in E_H, \ k \in [m]\}$,

- $lab_{H'}(u) = \begin{cases} lab_H(u) & \text{if } u \in E_H, \\ * & \text{otherwise} \end{cases}$ for $u \in V_{H'}$,

- $C = \{(ex/cr) \mid \exists \alpha \in T,\ k \in [m],\ i_1, \ldots, i_k, j_1, \ldots, j_k \in \mathbb{N}:$
  $lab(ex) = (\alpha, i_1 \ldots i_k),\ lab(cr) = (\alpha, j_1 \ldots j_k),$
  $\forall n \in [k]:\ (ex[n] \neq \Diamond \wedge cr[n] = n \wedge j_n = i_n)$ or
  $(ex[n] = \Diamond \wedge cr[n] \in V_{H'} \wedge (i_n, att_H(cr[n], j_n)) \in port)\}.$

Using this translation, $P' = \{X ::= \tau(H, port) \mid (X ::= (H, port)) \in P\}$ contains the productions of NG.

There is a one-to-one correspondence between the derivations in HG and the derivations in NG, and straightforward verification yields that $L(HG) = L(NG)$. The confluence of HG carries over to NG, and by construction, NG is remote-free.

$\square$

3.30 Corollary
$\mathcal{L}(\text{S-HH}) \subsetneq \mathcal{L}(\text{C-hNCE}_{\text{rf}}).$

Proof. By Theorem 3.29, $\mathcal{L}(\text{S-HH}) \subseteq \mathcal{L}(\text{C-hNCE}_{\text{rf}})$. By Theorem 7.5 in [CER93], there exists a hypergraph language $L \in \mathcal{L}(\text{HR}) \cap \mathcal{LH}$ which is not in $\mathcal{L}(\text{S-HH})$. (The set $plus(\text{DOTTEDTREES})$ discussed in Example 3.15 is such a language.) By Theorem 3.19, $L \in \mathcal{L}(\text{C-hNCE}_{\text{rf}})$. $\square$

To round off this section, we show that there are hypergraph languages in $\mathcal{L}(\text{C-hNCE}_{\text{rf}})$ which cannot be generated by either an HR grammar or an S-HH grammar.

3.31 Theorem (proper inclusion)
$(\mathcal{L}(\text{HR}) \cap \mathcal{LH}) \cup \mathcal{L}(\text{S-HH}) \subsetneq \mathcal{L}(\text{C-hNCE}_{\text{rf}}).$

Proof. Corollaries 3.21 and 3.30 state the inclusions.

We show that $L = plus(\text{DOTTEDTREES}) \cup \text{ADDCOMPLETE}$ is a C-hNCE$_{\text{rf}}$ language, but neither an HR nor an S-HH language.

L is a C-hNCE$_{rf}$ language. The set $plus(\text{DOTTEDTREES})$ is an HR language, see [CER93] and Example 3.15. This implies by Theorem 3.19 that it is also a C-hNCE$_{\text{rf}}$ language. Furthermore, ADDCOMPLETE is an S-HH language, see Example 3.25, and therefore a C-hNCE$_{\text{rf}}$ language by Theorem 3.29. Using the standard construction to unite context-free languages, see e.g. [HU79, Theorem 6.1], we obtain that $L = plus(\text{DOTTEDTREES}) \cup \text{ADDCOMPLETE}$ is a C-hNCE$_{\text{rf}}$ language.

L is not an HR language. Let the degree $deg_H(v)$ of a node v in a hypergraph H be the number of hyperedges incident to v in H, i.e. $deg_H(v) = \#eset_H(v)$. Moreover, define the minimum degree of a hypergraph H to be $mindeg(H) = \min\{deg_H(v) \mid v \in V_H\}$. Then Theorem 3.8 in [Hab92a, Chapter IV] states that the minimum degree of the hypergraphs in an HR language of simple hypergraphs is bounded.[3]

---

[3]The theorem is worded for simple graphs, but the proof for simple hypergraphs, with the notions of degree and minimum degree as given above, is literally the same and therefore not repeated here.

The language L, however, contains all 'complete graphs' and thus, for every $k \in \mathbb{N}$, a hypergraph H with $mindeg(H) \geq k$.

L is not an S-HH language. Suppose $L \in \mathcal{L}(\text{S-HH})$, and let HG be an S-HH grammar generating L. If $H \in L$ is a hypergraph in which the label a occurs, then H is a 'complete graph' where every node has a private hyperedge of rank 1 and labelled by a, and no other item of H is labelled by a. Thus, the handles induced by the a-labelled hyperedges are separated. Now construct an HH grammar HG' which differs from HG in that a is a nonterminal label, and HG' has in addition the production $a ::= \varepsilon$, where ε denotes the empty hypergraph (with ports). Then HG' is an S-HH grammar and generates the language $plus(\text{DOTTEDTREES}) \cup \{\varepsilon\}$. Using the standard construction for context-free languages, see e.g. [HU79, Theorem 4.3], an S-HH grammar HG'' can be obtained which is ε-production-free, i.e. which does not contain a production with right-hand side ε, such that

$$L(HG'') = L(HG') \smallsetminus \{\varepsilon\} = plus(\text{DOTTEDTREES}).$$

But this is a direct contradiction to Lemma 7.6 in [CER93], which implies that $plus(\text{DOTTEDTREES})$ can*not* be generated by an S-HH grammar. □

3.4 Limits of the Generative Power

In this section, first results on (hyper)graph languages which cannot be generated by a C-hNCE grammar are presented. As there is such a grammar for every finite set of hypergraphs, we need only consider infinite languages. We know already by Lemma 2.13 that (confluent) hNCE grammars cannot generate any hypergraph language of unbounded rank, such as the language HYPERSTARS of all hypergraphs consisting of one hyperedge linked to an arbitrary number of nodes as illustrated in Figure 3.10 or any language containing HYPERSTARS, such as $[\mathcal{H}]$. Moreover,

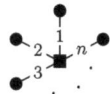

Figure 3.10: A hyperstar with n nodes

Theorem 3.10 implies that e.g. $[\mathcal{G}]$, the language CONNECTED of all connected graphs, or the language TOURNAMENTS of all tournaments (where a tournament is a graph with exactly one edge between every two distinct nodes) are not in $\mathcal{L}(\text{C-hNCE}_{rf})$, because they are not in $\mathcal{L}(\text{C-edNCE})$ (see [ER97, p. 63]). However, these graph languages might still be generated by a C-hNCE grammar making use of remote connection instructions, and Theorem 3.10 does not imply anything about *hyper*graph languages. In the following, an upper as well as a lower bound for the density of infinite C-hNCE languages is developed, where the term density refers to the number of hypergraphs in the language having (approximately) the same size, and the size of a hypergraph is measured in the number of its nodes. These bounds can then be used to prove that certain (hyper)graph languages cannot be generated by a C-hNCE grammar.

By eliminating Λ-productions as well as chain productions and making use of an enumeration argument (cf. [Wel84, Corollary 10]), one can show that for every C-hNCE language there is a certain upper bound on the number of hypergraphs in that language having the same size; see also [ER97, Theorem 1.3.30]. This allows to conclude that e.g. TOURNAMENTS and any language containing it, such as CONNECTED, $[\mathcal{G}]$, or the language HYPER$_k$ of all hypergraphs of rank at most k (for $k \geq 2$) cannot be generated by a C-hNCE grammar.

3.32 Definition (Λ-production)
A production $X ::= (R, C)$ of an hNCE grammar is a Λ-*production* if V_R is empty (there may be hyperedges or connection instructions, though). An hNCE grammar is *quasi Λ-production-free* if all Λ-productions have the start symbol as the left-hand side, and the start symbol does not label any node in the right-hand side of a production.

The elimination of Λ-productions is a standard operation on context-free string grammars, see e.g. [HU79, Theorem 4.3]. The only difference for C-hNCE grammars is that two Λ-productions with the same left-hand side need not be equal because the hyperedge sets or the connection relations of their right-hand sides can differ.

3.33 Lemma (quasi Λ-production-free normal form)
For every C-hNCE grammar, an equivalent C-hNCE grammar can be constructed which is quasi Λ-production-free.

Proof. Let $NG = (N, T, P, S)$ be a C-hNCE grammar. Compute the set M of all 'reachable Λ-productions' as follows:

- $M_0 = \{p \in P \mid p \text{ is a } \Lambda\text{-production}\}$,

- $M_{k+1} = M_k \cup \{X ::= (R, C)[v_1/(R_1, C_1)] \ldots [v_n/(R_n, C_n)] \mid$
  $(X ::= (R, C)) \in P, \text{ the nodes of } R \text{ are } v_1, \ldots, v_n,$
  $\text{and } (lab_R(v_i) ::= (R_i, C_i)) \in M_k \text{ for all } i \in [n]\}$,

- $M = M_k$ with $k \in \mathbb{N}$ minimal such that $M_k = M_{k+1}$.

Construct the hNCE grammar $NG' = (N', T, P', S')$ with $N' = N \cup \{S'\}$ (where S' is a new symbol not in $N \cup T$) and

$$
\begin{aligned}
P' = \quad & \{S' ::= (\bullet S , \emptyset)\} \\
& \cup \{S' ::= (R, C) \mid (S ::= (R, C)) \in M\} \\
& \cup (P \smallsetminus M_0) \\
& \cup \{X ::= (R, C)[v_1/(R_1, C_1)] \ldots [v_n/(R_n, C_n)] \mid \\
& \quad (X ::= (R, C)) \in P, \; \{v_1, \ldots, v_n\} \subsetneq V_R, \\
& \quad \text{and } (lab_R(v_i) ::= (R_i, C_i)) \in M \text{ for all } i \in [n]\}.
\end{aligned}
$$

By construction, the left-hand side of each of the Λ-productions is the start symbol S', which never occurs in the right-hand side of a production. By induction on the length of the derivation, each derivation $\bullet S' \Rightarrow^*_{P'} H'$ in NG' can be transformed into a derivation $\bullet S \Rightarrow^*_P H$ in NG as follows: remove the first step if it is $\bullet S' \Rightarrow \bullet S$ resp. relabel the node in the start hypergraph with S if it is not, and in the rest of the derivation, whenever any production $p \in P' \smallsetminus P$ is applied, apply instead the combination of productions of NG which gave rise to p. By the associativity of NG and the construction of P', it follows that $H = H'$. Conversely, the confluence of NG implies that the steps in each of its derivations can be re-ordered so that by associativity (and relabelling the node in the start hypergraph with S' or adding a first derivation step $\bullet S' \Rightarrow \bullet S$) a derivation in NG' can be constructed which generates the same hypergraph. In other words, NG and NG' are equivalent. Finally, the confluence of NG implies the confluence of NG'.   □

3.34 Definition (chain production)
A production $p = (X ::= (R, C))$ of an hNCE grammar is a *chain production* if R has a single node with a nonterminal label (R may have hyperedges).

The removal of chain productions is a further standard operation on context-free string grammars (see e.g. [HU79, Theorem 4.4]) and can be performed analogously for C-hNCE grammars.

3.35 Lemma (chain-production-free normal form)
For every C-hNCE grammar NG, an equivalent C-hNCE grammar NG' without chain productions can be constructed. If NG is quasi Λ-production-free, then so is NG'.

Proof. Let $NG = (N, T, P, S)$ be a C-hNCE grammar; NG can be assumed to be quasi Λ-production-free by Lemma 3.33. Compute the set M of all 'reachable chain productions' as follows (where v is an arbitrary but fixed node, the unique node in the right-hand side hypergraph of every chain production in M):

- $M_0 = \{(X ::= (R, C)) \in copy(P) \mid V_R = \{v\}\}$,

- $M_{k+1} = M_k \cup \{X ::= (R, C) \mid$

$$\exists (X ::= (R_1, C_1)), (lab_{R_1}(v) ::= (R_2, C_2)) \in M_k,$$
$$(R_1', C_1') \in [(R_1, C_1)] \text{ with } V_{R_1'} = \{v'\} \text{ and } v' \neq v,$$
$$\text{and } (R, C) = (R_1', C_1')[v'/(R_2, C_2)]\},$$

- $M = M_k$ with $k \in \mathbb{N}$ minimal such that $M_k = M_{k+1}$.

Now construct the hNCE grammar $NG' = (N, T, P', S)$ with

$$P' = \{p \in P \mid p \text{ is not a chain production}\} \cup$$
$$\{X ::= (R, C) \mid \exists (X ::= (R_1, C_1)) \in M,$$
$$(lab_{R_1}(v) ::= (R_2, C_2)) \in P \text{ is not a chain production},$$
$$(R_1', C_1') \in [(R_1, C_1)] \text{ with } V_{R_1'} = \{v'\} \text{ and } v' \notin V_{R_2},$$
$$\text{and } (R, C) = (R_1', C_1')[v'/(R_2, C_2)]\}.$$

By construction, NG' does not contain chain productions and is quasi Λ-production free. The equivalence of NG and NG' as well as the confluence of NG' can be shown similarly as in the proof of Lemma 3.33, using the confluence and associativity of NG. $\qquad\square$

Now we are ready to show an upper bound for the density of C-hNCE languages, which is the same as for C-edNCE languages (see [Wel84, Corollary 10] and also [ER97, Theorem 1.3.30]).

3.36 Theorem (upper bound for density)
For every C-hNCE language L there is a constant c such that for every $n \geq 1$,
$$\#\{[H] \in L \mid \#V_H = n\} \leq 2^{c \cdot n}.$$

Proof. Let NG be a C-hNCE grammar which generates L. By Lemmas 3.33 and 3.35, we may assume that NG is quasi Λ-production-free and does not contain chain productions. Then any derivation generating a hypergraph $[H] \in L$ with $n \geq 1$ nodes is of length less than $2n$. Assume an arbitrary but fixed order on the nonterminal nodes in the right-hand sides of the productions in NG, and consider, for each hypergraph $[H] \in L$, a leftmost derivation[4] in NG which generates H. Up to isomorphism, such a derivation—and with it $[H]$—is completely specified by the sequence of productions in NG as they are applied in the sequence of derivation steps. As a derivation generating a terminal hypergraph cannot be a proper prefix of another derivation, there are at most d^{2n} such production sequences of length up to $2n$ which specify the derivation of a hypergraph in L with up to n nodes, where d is the number of productions in NG. Thus, d^{2n} (or $2^{c \cdot n}$ with c suitably chosen) is an upper bound for the number of hypergraphs in L with $n \geq 1$ nodes. $\qquad\square$

---

[4]The assumed order induces an order on the nonterminal nodes of the sentential forms of NG, and a derivation is leftmost if in each step the rewritten node is the least of all nodes rewritten in the rest of the derivation. For a precise definition see Definition 4.11.

As stated in [ER97, p. 63], graph languages like $[\mathcal{G}]$, CONNECTED, or TOUR-
NAMENTS cannot be C-edNCE languages because in each of them, the number of
graphs having the same size grows faster than $2^{c \cdot n}$. Consequently, Theorem 3.36
implies that they cannot be generated by a C-hNCE grammar either, even if the
grammar is not remote-free. Moreover, the same holds for any hypergraph language
containing one of these languages, such as the language HYPER$_k$ consisting of all
hypergraphs of rank at most k (for $k \geq 2$), as well as for hypergraph variants of
these languages, such as *plus_one*(TOURNAMENTS) consisting of all 'tournaments'
with one additional node made incident to every edge by a third tentacle (see
Figure 3.11).

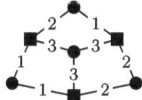

Figure 3.11: A hypergraph in *plus_one*(TOURNAMENTS); the additional node is the
one in the middle

For C-hNCE languages, a generalisation of Parikh's theorem for context-free
string languages [Par66] can be shown analogously to the case of hyperedge-replace-
ment languages (see [Hab92a, Theorem 4.3 in Chapter IV] and also [DHK97, Theorem 2.4.7]). This is possible due to the non-blocking normal form of C-hNCE
grammars (Theorem 3.7). A consequence of Parikh's theorem for C-hNCE languages is that the size of the hypergraphs in such a language can grow at most
linearly, which yields a second criterion to prove that a hypergraph language cannot be generated by C-hNCE grammars.

3.37 Definition (Parikh mapping)
Let $T = \{a_1, \ldots, a_n\}$ be a finite set of symbols and $\psi \colon [\mathcal{H}_T] \to \mathbb{N}^n$ the mapping
given by $\psi([H]) = (\#_{a_1}([H]), \ldots, \#_{a_n}([H]))$, where $\#_{a_i}([H])$ denotes the number
of a_i-labelled nodes for $[H] \in [\mathcal{H}_T]$. Then ψ is called a *Parikh mapping*. For every
hypergraph language $L \subseteq [\mathcal{H}_T]$, $\psi(L)$ denotes the set $\psi(L) = \{\psi([H]) \mid [H] \in L\}$.

3.38 Definition (semilinear set)
A set $S \subseteq \mathbb{N}^n$ is *linear* if S is of the form

$$S = \{x_0 + \sum_{i=1}^{k} c_i \cdot x_i \mid c_1, \ldots, c_k \in \mathbb{N}\},$$

where $k \geq 1$ and $x_0, \ldots, x_k \in \mathbb{N}^n$. A subset of $\mathbb{N}^n$ is *semilinear* if it is the union of
finitely many linear sets.

Now we can generalise Parikh's theorem to C-hNCE languages analogously to [DHK97, Theorem 2.4.7].

3.39 Theorem (Parikh's theorem)
For all C-hNCE languages L and each Parikh mapping ψ, the set $\psi(L)$ is semilinear.

Proof. Let L be a C-hNCE language and $NG = (N, T, P, S)$ a C-hNCE grammar generating L. By Theorem 3.7, we may assume that NG is non-blocking.

Let $string \colon \mathcal{H}_{N \cup T} \to (N \cup T)^*$ be a mapping assigning a string $string(H) = lab_H(v_1) \ldots lab_H(v_n)$ to a hypergraph H with $V_H = \{v_1, \ldots, v_n\}$ (where the order on the nodes is arbitrary), and construct the context-free string grammar $Gr = (N, T, P', S)$ with $P' = \{X ::= string(R) \mid (X ::= (R, C)) \in P\}$.

As NG is non-blocking, a sentential form H of NG has only terminally labelled nodes if and only if $[H]$ is in $L(NG)$ if and only if $string(H)$ is in $L(Gr)$. Therefore, $\psi(L(NG)) = \psi_{string}(L(Gr))$, where ψ_{string} denotes the usual Parikh mapping for string languages. By Parikh's theorem for context-free string languages, the set $\psi_{string}(L(Gr))$ is semilinear, which implies the same for $\psi(L(NG))$. □

Theorem 3.39 implies that the size of the hypergraphs in a C-hNCE language grows at most linearly.

3.40 Corollary (lower bound for density)
For every infinite C-hNCE language L, there are constants $c, c' \in \mathbb{N}$ such that L contains for all $n \in \mathbb{N}$ a hypergraph with $c' + n \cdot c$ nodes.

Proof. Let $L \in [\mathcal{H}_T]$ be an infinite C-hNCE language and ψ a Parikh mapping on $[\mathcal{H}_T]$. By Theorem 3.39, $\psi(L)$ is a finite union of linear sets. Let

$$S = \{x_0 + \sum_{i=1}^{k} c_i \cdot x_i \mid c_1, \ldots, c_k \in \mathbb{N}\}$$

be one of these sets and $i \in [k]$ such that x_i is not the 0-vector. Then S—and with it L—contains for all $c_i \in \mathbb{N}$ the vector $x_0 + c_i \cdot x_i$ (choose $c_j = 0$ for all $j \neq i$), i.e. L contains for all $c_i \in \mathbb{N}$ a hypergraph with $x_0[l] + c_i \cdot x_i[l]$ many a_l-labelled nodes for all $l \in [\#T]$ (where $x[l]$ denotes the lth entry of the vector x). Consequently, there is for each $n \in \mathbb{N}$ a hypergraph in L with $c' + n \cdot c$ nodes, where c' resp. c is the sum of all entries in x_0 resp. x_i. □

With Corollary 3.40, we have a second criterion to prove that certain (hyper)graph languages are not in $\mathcal{L}$(C-hNCE): For example, neither EXP-STARS containing all stars with exponentially many rays as shown in Figure 3.12(a) (the size grows exponentially) nor GRID^2 containing all square grids as shown in Figure 3.12(b) or *plus*(COMPLETE) obtained by adding to each edge of a complete graph a new private node as shown in Figure 3.12(c) (the size grows quadratically) can be C-hNCE languages.

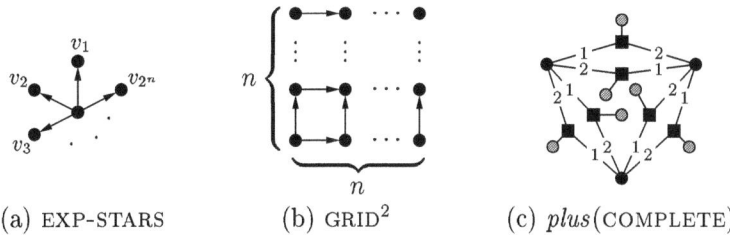

(a) EXP-STARS   (b) GRID^2 (c) *plus*(COMPLETE)

Figure 3.12: Hypergraph languages where the number of nodes grows faster than linear

3.5 Concluding Remarks

C-hNCE grammars currently represent the most general known context-free hyper-graph-rewriting approach, properly generalising both hyperedge and separated handle rewriting [Kle99]. Nevertheless, the graph-generating power of remote-free C-hNCE grammars coincides with that of C-edNCE grammars [Kle]. Moreover, we conjecture that the use of remote connection instructions does not increase the generative power of C-hNCE grammars. The diagram of Figure 3.13 summarises these relationships; if the conjecture above is correct, the grey area is empty.

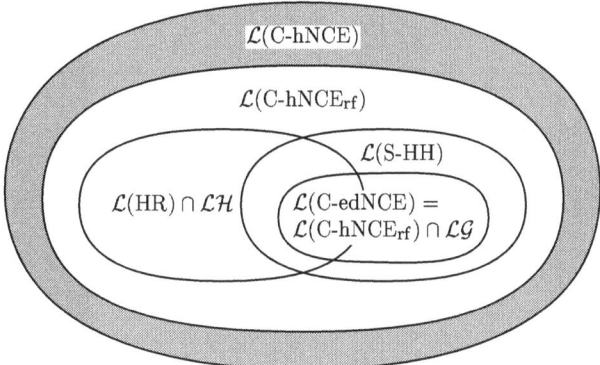

Figure 3.13: Comparison of generative power

Kim and Jeong introduced S-HRNCE grammars as a second separated handle rewriting approach in [KJ99]. They are based on node and hyperedge labelled multiple hypergraphs with undirected hyperedges, i.e. a hyperedge has a *set* instead of a *sequence* of attachment nodes, and without isolated nodes or hyperedges with empty attachment set. Rewriting a handle defined by a hyperedge e in a hypergraph H_1 with a hypergraph H_2 consists of *removing* e together with all incident nodes and all tentacles gripping to these nodes, yielding the remainder H_1^- of H_1; *adding* H_2 to H_1^-; connecting H_2 and H_1^- according to some connection relation which contains triples (v, a, B) saying that if in H_1 an a-labelled node was incident both to e and a B-labelled hyperedge e' (distinct from e), then the node v of H_2 is added to the attachment set of e'; and *deleting* from the resulting hypergraph all those hyperedges that have an empty attachment set. S-HRNCE grammars are not naturally context-free like S-HH grammars (but they do have a context-free normal form [KJ99, Theorem 5.25]). Moreover, S-HRNCE grammars do not offer a way to control the rank of the hyperedges, so that hypergraph languages of unbounded rank can be generated. Due to the important differences in the underlying hyper-graph model and the properties of the rewriting mechanism, the generative power of S-HRNCE grammars is not related to that of C-hNCE grammars in this thesis.

4

Decidability of Context-freeness for hNCE Grammars

Context-free, i.e. confluent, hNCE languages have been established in the preceding chapter (see also [Kle99]) as the most general known class of hypergraph languages. As confluence is a dynamic property which cannot be verified in a single glance at a grammar, it is therefore of interest to know—or to know how to find out—whether a given hNCE grammar is confluent. In this chapter, the decidability of confluence for hNCE grammars is studied.

For edNCE grammars, Kaul answered this question positively in [Kau85]. Given an edNCE grammar, it suffices to enumerate all (abstract) subgraphs of sentential forms induced by two nonterminal nodes, the so-called critical candidates, and test whether they satisfy the confluence condition. The number of possible critical candidates is exponential in the size of the alphabets in the grammar, and a new candidate can be computed from an old one (or from a one-node subgraph of a sentential form) in one derivation step. Thus, the actual candidates are reached from the axiom after at most $n^2 \cdot 2^{2n} + n$ derivation steps, where n is the size of the alphabets.

For hNCE grammars, confluence is decidable, too [Kle96]. However, the high flexibility for the generation of embedding hyperedges leads to significantly more complex substructures of sentential forms which can be used as critical candidates, and in general more than one derivation step is needed to compute a new critical candidate from an old one. All in all, the algorithm proposed here is doubly exponential in the size of the grammar.

In Section 4.1, a notion of critical candidate for hNCE grammars is developed. The confluence test is presented in Section 4.2, and its correctness and termination are proved in Section 4.3. As an alternative to the highly complex confluence test, the notion of static confluence is briefly discussed in Section 4.4.

4.1   Critical Candidates

For edNCE grammars, confluence can be decided as follows: Suppose that the grammar is not confluent, i.e. there is a sentential form G containing nodes v_1 and v_2 which are rewritten with productions p_1 and p_2, respectively, yielding graphs G_{12} and G_{21} which are not the same, where G_{ij} denotes the graph derived from G by first applying p_i to v_i, then p_j to v_j. This can only be if there is an edge in G incident to both v_1 and v_2 which leads to an embedding edge e' in, say, G_{12}, but e' does not belong to G_{21}. Thus it suffices to compute all (abstract) two-node subgraphs of sentential forms, so-called critical candidates, and use them to locally test for confluence. If there are λ (nonterminal or terminal) labels in the grammar, then there are at most $\lambda^2 \cdot 2^{2\lambda}$ non-isomorphic two-node graphs: each node has a label (which accounts for the first factor), and the edges are a subset of maximally λ edges from the first to the second node and their λ inverse edges (the second factor). As a new critical candidate can be obtained either as a subgraph of the right-hand side of a production or by rewriting one of the nodes of a previously reached critical candidate, this implies that if M_i contains all abstract one- and two-node subgraphs of the sentential forms derived from the start graph in up to i steps for $i \in \mathbb{N}$ (in particular $M_0 = \{[\bullet S]\}$), then M_μ with $\mu = \lambda^2 \cdot 2^{2\lambda} + \lambda$ contains all reachable critical candidates. Moreover, M_{i+1} can be computed directly from M_i, i.e. without having to consider arbitrarily large sentential forms, and testing a critical candidate for confluence can be done in constant time. Therefore, confluence is decidable for edNCE grammars in exponential time.

In this section, a structure which can play the role of a critical candidate for hNCE grammars is determined. For this, the notion of confluence has to be broken down to a condition for (parts of) a hypergraph.

4.1 Definition (local confluence)
Let $NG = (N, T, P, S)$ be an hNCE grammar. A hypergraph H over $N \cup T$ is NG-confluent with respect to $U \subseteq V_H$ if for all distinct nonterminal nodes $v_1, v_2 \in U$ and all production copies $lab_H(v_1) ::= (R_1, C_1)$ and $lab_H(v_2) ::= (R_2, C_2)$ in $copy(P)$ such that H, (R_1, C_1), and (R_2, C_2) are mutually disjoint, the order in which the substitutions $[v_1/(R_1, C_1)]$ and $[v_2/(R_2, C_2)]$ are performed on H has no influence on the subhypergraph of the resulting hypergraph which is induced by $U' := (U \setminus \{v_1, v_2\}) \cup V_{R_1} \cup V_{R_2}$, i.e.

$$H[v_1/(R_1, C_1)][v_2/(R_2, C_2)]\,|_{U'} = H[v_2/(R_2, C_2)][v_1/(R_1, C_1)]\,|_{U'}.$$

The hypergraph H is NG-confluent if it is NG-confluent with respect to V_H.

4.2 Lemma (local vs. global confluence)
(1) An hNCE grammar NG is confluent if and only if each of its sentential forms is NG-confluent.
(2) Let $NG = (N, T, P, S)$ be an hNCE grammar and $H \in \mathcal{H}_{H \cup T}$. The hypergraph H is NG-confluent if and only if for all hyperedges $e \in E_H$, H is NG-confluent with respect to $vset_H(e)$.

Proof. (1) This follows immediately from the definitions.
(2) If H is *NG*-confluent, then it is so with respect to each subset of V_H, too.

If H is not *NG*-confluent, then there are nonterminal nodes $v_1, v_2 \in V_H$ and production copies $(lab_H(v_1) ::= (R_1, C_1))$, $(lab_H(v_2) ::= (R_2, C_2)) \in copy(P)$ such that H, (R_1, C_1), and (R_2, C_2) are pairwise disjoint and

$$H_{12} = H[v_1/(R_1, C_1)][v_2/(R_2, C_2)] \neq H[v_2/(R_2, C_2)][v_1/(R_1, C_1)] = H_{21}.$$

The definition of node substitution implies that H_{12} and H_{21} have the same nodes, so they can only differ in some hyperedge e which is in, say, H_{12} but not in H_{21}. Thus, the subhypergraphs of H_{12} and H_{21} induced by the nodes incident to e in H_{12} differ, too. As H_{12} and H_{21} both contain the hyperedges of R_1, R_2, and the hyperedges of H not incident to v_1 or v_2, e must be an embedding hyperedge. This means that there are a hyperedge $e'' \in E_H$ and connection instructions $coin_1 \in C_1$, $coin_2 \in C_2$ such that $coin_1$ transformes e'' into a hyperedge e' in $H[v_1/(R_1, C_1)]$ and $coin_2$ transformes e' into e. Thus the nodes incident to e form a subset of $(vset_H(e'') \setminus \{v_1, v_2\}) \cup V_{R_1} \cup V_{R_2}$, and H is not *NG*-confluent with respect to $vset_H(e'')$. $\square$

Unlike the confluence test for edNCE grammars, the (abstract) subhypergraphs of the sentential forms induced by the incident nodes of some hyperedge do not provide correct structures to locally test confluence for hNCE grammars.

4.3 Example
Let $NG = (N, T, P, S)$ be an hNCE grammar where P consists of the production $p = (X ::= (\bullet a, \{(\beta, X \lozenge / \alpha, 2)\}))$. The application of p to two adjacent nodes of the hypergraph $H \in \mathcal{H}_{H \cup T}$ in Figure 4.1 shows that this particular hypergraph is *NG*-confluent. However, the subhypergraph of H induced by the two rewritten nodes is not *NG*-confluent. ∎

To define a structure which allows for a local confluence test of hNCE grammars, the notion of a *locale* will be used. If H is a hypergraph and U a subset of its nodes, then the locale of U in H consists roughly of the subhypergraph of H induced by U plus all hyperedges of H which have at least one incident node in U. The nodes incident to such a hyperedge, but not belonging to U, are then glued together according to their labels. Figure 4.2 shows a sketch of this construction.

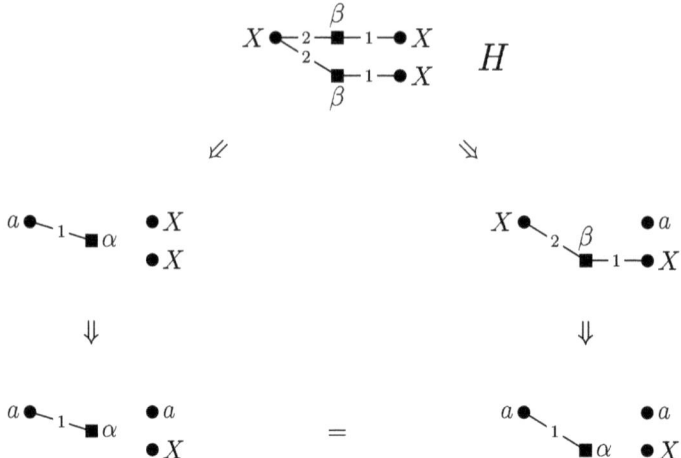

Figure 4.1: An *NG*-confluent hypergraph with non-confluent subhypergraph

4.4 Definition (locale)

Let $NG = (N, T, P, S)$ be an hNCE grammar, $H \in \mathcal{H}_{N \cup T}$, and $U \subseteq V_H$. The *locale* of U in H is the hypergraph $loc(H, U) = (V, E, lab)$ with

- $V = U \uplus (N \cup T)$,

- $E = \{(\alpha, u_1 \dots u_n) \mid (\alpha, v_1 \dots v_n) \in E_H, \exists i \in [n] : v_i \in U,$
  $\forall i \in [n] : (v_i \in U \wedge u_i = v_i) \vee (v_i \notin U \wedge u_i = lab_H(v_i))\}$, and

- $lab(u) = \begin{cases} lab_H(u) & \text{if } u \in U \\ u & \text{if } u \in N \cup T \end{cases}$ for all $u \in V$.

The set U is the *centre* of $loc(H, U)$, denoted by $centre(loc(H, U))$.

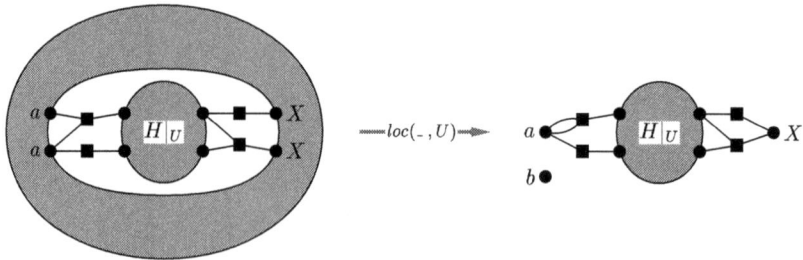

Figure 4.2: Constructing the locale $loc(H, U)$ of a hypergraph over $\{a, b, X\}$

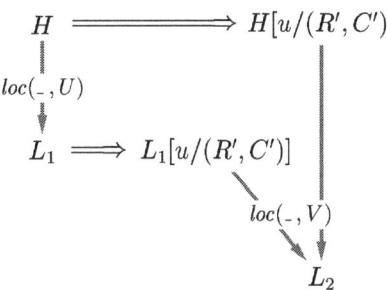

Figure 4.3: Locales can monitor a substitution

It is immediate that $H|_U = loc(H,U)|_U$ for all hypergraphs H and all node sets $U \subseteq V_H$. Moreover, $loc(loc(H,U),U') = loc(H,U')$ for all $U' \subseteq U$.

In order to show that locales support a local confluence test, the following lemma is useful which says that the effects of replacing a centre node in a locale are monitored correctly on the limited scope of the centre of the locale. The situation is illustrated in Figure 4.3, where the construction of a locale $loc(\_,U)$ is meant as a function mapping any hypergraph H which contains the nodes in U to the locale $loc(H,U)$.

4.5 Lemma (monitoring confluence in locale)
Let $NG = (N,T,P,S)$ be an hNCE grammar, $H \in \mathcal{H}_{NUT}$, $U \subseteq V_H$, $u \in U$, $lab_H(u) ::= (R,C)$ a production in $copy(P)$ with (R,C) disjoint from H, and $V = (U \smallsetminus \{u\}) \cup V_R$.
  Then $loc(loc(H,U)[u/(R,C)], V) = loc(H[u/(R,C)], V)$.

Proof. Clearly, the two locales have the same nodes.

All hyperedges in $loc(H,U)$ which are not incident to u and all hyperedges in R belong to both locales, too, and so do all embedding hyperedges which are obtained from some hyperedge $e \in E_H$ with $vset_H(e) \subseteq U$.

If e' is a hyperedge in $H[u/(R,C)]$ obtained from a hyperedge $e \in E_H$ with $u \in vset_H(e) \not\subseteq U$ via a connection instruction *coin*, then there is a hyperedge $\bar{e}$ in $loc(H,U)$ which has the same incidences to nodes in U and which also fits the existence part of *coin*. Thus *coin* transforms $\bar{e}$ into a hyperedge $\bar{e}'$ in $loc(H,U)[u/(R,C)]$ which has the same incidences to nodes in V as the hyperedge e', and which is incident to $lab_{H[u/(R,C)]}(v)$ where e' is incident to $v \notin V$.

Conversely, if there is a hyperedge $\bar{e}'$ in $loc(H,U)[u/(R,C)]$ obtained from a hyperedge $\bar{e}$ in $loc(H,U)$ with $u \in vset_{loc(H,U)}(\bar{e}) \not\subseteq U$ via a connection instruction *coin*, then there is a corresponding hyperedge e in H which is transformed by *coin* into a hyperedge e' in $H[u/(R,C)]$, and $\bar{e}'$ and e' are related as above.

In both cases, e' and $\bar{e}'$ are mapped to the same hyperedge by $loc(\_,V)$. $\square$

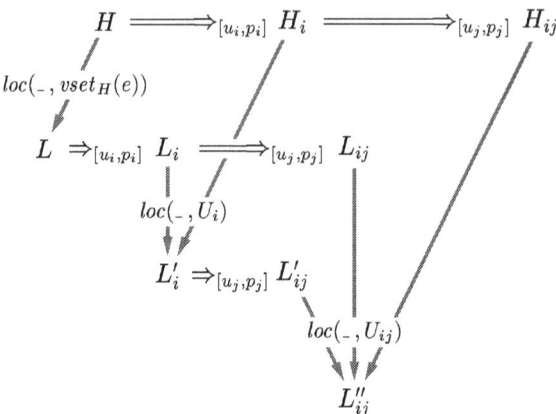

Figure 4.4: Monitoring two consecutive substitutions

4.6 Theorem (local confluence in locale)
(1) Let $NG = (N, T, P, S)$ be an hNCE grammar, $H \in S(NG)$, and $e \in E_H$.

The hypergraph H is NG-confluent with respect to $vset_H(e)$ if and only if $loc(H, vset_H(e))$ is NG-confluent with respect to $vset_H(e)$.

(2) An hNCE grammar $NG = (N, T, P, S)$ is confluent if and only if for all sentential forms $H \in S(NG)$ and all hyperedges $e \in E_H$, $loc(H, vset_H(e))$ is NG-confluent with respect to $vset_H(e)$.

Proof. (1) For $i = 1, 2$, let u_i be a node in $vset_H(e)$ and $lab_H(u_i) ::= (R_i, C_i)$ a production in $copy(P)$ such that H, (R_1, C_1), and (R_2, C_2) are pairwise disjoint. Consider the situation shown in Figure 4.4, for hypergraphs $K_i = K[u_i/(R_i, C_i)]$ and $K_{ij} = K[u_i/(R_i, C_i)][u_j/(R_j, C_j)]$ and node sets $U_i = (vset_H(e) \smallsetminus \{u_i\}) \cup V_{R_i}$ and $U_{ij} = (vset_H(e) \smallsetminus \{u_i, u_j\}) \cup V_{R_i} \cup V_{R_j} = U_{ji}$, where $K \in \{H, L, L', L''\}$ and $i, j \in \{1, 2\}$ with $i \neq j$. By triple usage of Lemma 4.5 we have $L_{ij}|_{U_{ij}} = L''_{ij}|_{U_{ij}} = H_{ij}|_{U_{ij}}$, which in turn implies $L_{12}|_{U_{12}} = L_{21}|_{U_{21}}$ if and only if $H_{12}|_{U_{12}} = H_{21}|_{U_{21}}$.
(2) This follows immediately from (1) and Lemma 4.2.   □

Part (2) of Theorem 4.6 justifies using certain abstract locales as critical candidates for hNCE grammars.

4.7 Definition (critical candidate)
Let $NG = (N, T, P, S)$ be an hNCE grammar. Two locales L, L' are isomorphic if there is a hypergraph isomorphism $f \colon L \to L'$ mapping $centre(L)$ onto $centre(L')$ (and thus extending the identity on $N \cup T$). Then a *critical candidate* of NG is an abstract locale $[loc(H, U)]$ where H is a sentential form of NG and $U = vset_H(e)$ for some $e \in E_H$ incident to two (or more) nonterminal nodes.

As in the case of edNCE grammars, the critical candidates can be computed as part of a slightly larger set of abstract locales.

4.8 Definition (setting)
Let $NG = (N, T, P, S)$ be an hNCE grammar. A *setting* of NG is an abstract locale $[loc(H, U)]$ where H is a sentential form of NG, $U = \{u\}$ for some $u \in V_H$ or $U = vset_H(e)$ for some $e \in E_H$, and U contains at least one nonterminally labelled node (such a set U is called a *setting centre*). The set of settings of NG is denoted by $Sett(NG)$.

4.9 Lemma (upper bound for the number of settings)
Let $NG - (N, T, P, S)$ be an hNCE grammar, $\kappa = rank(S(NG)) \geq 1$, and $\lambda = \#(N \cup T)$. Then the size of $Sett(NG)$ is bounded by

$$\sigma = \kappa \cdot \lambda^\kappa \cdot 2^{(\kappa+1)\cdot\lambda\cdot(\kappa+\lambda)^\kappa}.$$

Proof. A locale in a setting has between 1 and κ centre nodes, which means between $1 + \lambda$ and $\kappa + \lambda$ nodes in total. In a hypergraph with j nodes, there are up to $\lambda \cdot j^i$ hyperedges of rank i, where $i \in [\kappa] \cup \{0\}$. Thus the maximal number of hyperedges in a hypergraph with j nodes is $\sum_{i \in [\kappa] \cup \{0\}} \lambda \cdot j^i$, which is bounded by $(\kappa + 1) \cdot \lambda \cdot j^\kappa$. As the number of centre nodes can vary from 1 to κ, all centre nodes of a locale have some node label, and the hyperedges of a particular hypergraph form a subset of all possible hyperedges, σ provides an upper bound for the number of distinct settings. $\square$

4.2   A Confluence Test

Analogously to the edNCE case, the basic idea of the confluence test presented in this section is to enumerate the settings of a given hNCE grammar NG and test them for NG-confluence with respect to their centre. The first setting is obviously the one constructed from the axiom of the grammar, with the unique node of the axiom as centre. Computing a new setting from an old one is, however, more difficult than for edNCE grammars. Given a setting $[loc(H, U)]$ of some hNCE grammar NG, a new setting can be obtained by performing a rewriting step $loc(H, U) \Rightarrow_{[u,p]} L$ with $u \in U$, choosing a new setting centre among the nodes of $U \smallsetminus \{u\}$ and the descendants of u, and constructing the corresponding setting. But just repeating these steps means that probably some settings are not computed.

4.10 Example
Let $NG = (N, T, P, S)$ be an hNCE grammar, let P contain the production $p = (X ::= (\bullet a, \{(\beta, X\Diamond/\alpha, 2)\}))$, and let the hypergraph H in Figure 4.5 be a sentential form of NG. Three settings can be constructed from H by choosing as centre

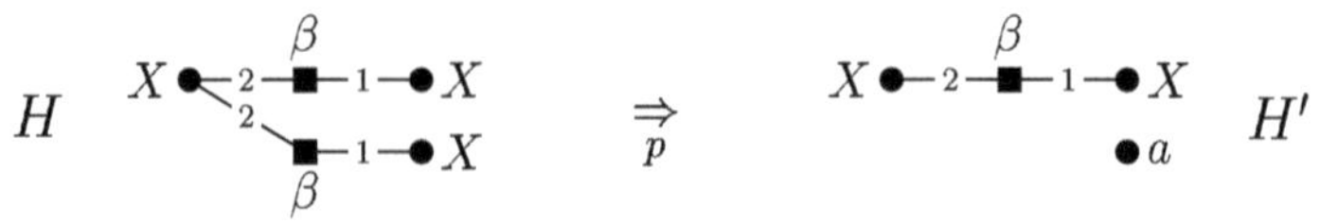

$$H \quad X \bullet \overset{\beta}{\underset{2}{\leftarrow 2 - \blacksquare - 1 \rightarrow}} \bullet X \qquad \Rightarrow \atop p \qquad X \bullet - 2 - \blacksquare - 1 - \bullet X \quad H'$$

Figure 4.5: New settings can be reached by rewriting non-centre nodes

the node on the left, or either of the nodes on the right, or the incident nodes of either of the hyperedges. However, a forth setting can be obtained if e.g. the nodes incident to the upper hyperedge are chosen as the new centre, and then the node not belonging to the centre is rewritten yielding the hypergraph H' before the setting is constructed.   ∎

So, the first idea has to be modified, for example as follows. For an hNCE grammar NG, the settings reachable from a setting $[loc(H, U)]$ of NG can be computed by iterating the following nondeterministic steps:

1. Rewrite a node $u \in U$ with a production $X ::= (R, C)$ of NG.

2. Choose a new setting centre $U' \subseteq (U \smallsetminus \{u\}) \cup V_R$.

3. For each node in $(U \smallsetminus \{u\}) \cup V_R$, but not in U', perform some derivation in which only descendants of that node are rewritten.

4. Construct from the resulting hypergraph the setting with centre U'.

In general, there is no bound on the length of the derivations done in step 3, so that this algorithm may not terminate. But making use of the associativity of the grammar, each of these derivations can be equivalently replaced with one derivation step in which the node v is rewritten. Considering that this derivation step contributes to the setting to be computed only with hyperedges of rank 0 and with embedding hyperedges incident to at least one centre node, each hypergraph with embedding substituted for such a node v can be reduced to this relevant information. As there are only finitely many of these reduced hypergraphs with embedding, an according modification to the algorithm will guarantee termination.

It will be shown in the next section that it is not necessary to consider all possible derivations for step 3; instead, leftmost derivations are sufficient. To define this notion, an hNCE grammar has to be equipped with an (arbitrary) order on the nonterminal nodes in the right-hand sides of its productions, cf. [ER97]. In contrast to that reference, we do not require that the first nonterminal node of the given order has to be rewritten in a leftmost derivation, but if it is rewritten at all in the derivation, then this has to be done in the first step. It is immediate that in a confluent hNCE grammar, every (subderivation of a) derivation of a sentential form can be transformed into a leftmost (sub)derivation which generates the same hypergraph.

4.11 Definition (leftmost derivation)

Let $NG = (N, T, P, S)$ be an hNCE grammar. A *nonterminal-ordered* hypergraph with embedding over $N \cup T$ is a hypergraph with embedding (H, C) over $N \cup T$ together with a linear order on its nonterminal nodes. The grammar NG is nonterminal-ordered if the right-hand sides of all its productions are. If (H, C) and (H', C') are disjoint hypergraphs with embedding over $N \cup T$, $(u_1, \ldots, u_m)$ is a linear order on the nonterminal nodes of H, and $(v_1, \ldots, v_n)$ is a linear order on the nonterminal nodes of H', then for $i \in [m]$ the derived order on the nonterminal nodes of $(H, C)[u_i/(H', C')]$ is $(u_1, \ldots, u_{i-1}, v_1, \ldots, v_n, u_{i+1}, \ldots, u_m)$.

For a nonterminal-ordered hypergraph with embedding (H, C) over $N \cup T$, a derivation

$$(H, C) \Rightarrow_{[v_1, p_1]} (H_1, C_1) \Rightarrow \cdots \Rightarrow_{[v_n, p_n]} (H_n, C_n)$$

is *leftmost*, denoted $(H, C) \Rightarrow_P^n (H_n, C_n)$, if for all $i \in [n]$ either v_1 is the ancestor of v_i in H or v_1 is prior to that ancestor in the order of (H, C), and

$$(H_1, C_1) \Rightarrow \cdots \Rightarrow_{[v_n, p_n]} (H_n, C_n)$$

is leftmost.

Reduced hypergraphs with embedding are defined formally as follows.

4.12 Definition (reduced hypergraph with embedding)

Let $NG = (N, T, P, S)$ be an hNCE grammar. Define, for a hypergraph with embedding (H, C) over $N \cup T$, the *reduced* hypergraph with embedding $red(H, C) = (\Sigma_H, C_H)$ with $V_{\Sigma_H} = N \cup T$, $E_{\Sigma_H} = \{e \in E_H \mid rank_H(e) = 0\}$, $lab_{\Sigma_H} = id$, and C_H consisting of the connection instructions of C, but with the nodes in the creation parts replaced by their respective labels in H.

For each nonterminal symbol X, we can compute the set $Red(X)$ of reduced hypergraphs with embedding which are obtained by reducing the hypergraphs with embedding leftmost derivable from the right-hand side of a production with X as left-hand side as follows.

4.13 Definition ($Red(X)$)

Let $NG = (N, T, P, S)$ be a nonterminal-ordered hNCE grammar. For all $X \in N$, compute in parallel sets $Red(X)_k$ of reduced hypergraphs with embedding as follows.

- $Red(X)_0 = \emptyset$ and

- $Red(X)_{k+1} = \quad Red(X)_k$

  $\cup \{ red(R_n, C_n) \mid \exists\, (X ::= (R, C)) \in P,$

  $\quad (R, C) \Rightarrow_{[v_1, p_1]} (R_1, C_1) \Rightarrow \cdots \Rightarrow_{[v_n, p_n]} (R_n, C_n)$

  $\quad$ where $(v_1, \ldots, v_n)$ are n nonterminal nodes of R

  $\quad$ in the fixed order, for some $n \in \mathbb{N}$,

  $\quad$ and for all $i \in [n]$: $p_i = (lab_R(v_i) ::= (R'_i, C'_i))$

  $\quad$ with $(R'_i, C'_i) \in Red_k(lab_R(v_i))\}$.

For $k \in \mathbb{N}$ such that $Red(X)_k = Red(X)_{k+l}$ for all $X \in N$ and $l \in \mathbb{N}$, $Red(X)_k$ will be denoted by $Red(X)$.

Now we are ready to compute (all) settings of a (confluent) hNCE grammar, using the idea discussed after Example 4.10.

4.14 Definition ($Sett$)
Let $NG = (N, T, P, S)$ be a nonterminal-ordered hNCE grammar, and let the sets $Red(X)$ for $X \in N$ be computed as above. Moreover, let $\bullet S$ be the axiom of NG and x the node of $\bullet S$.
Define, for $k \in \mathbb{N}$, sets $Sett_k$ as follows.

- $Sett_0 = \{[loc(\bullet S, \{x\})]\} \quad$ and

- $Sett_{k+1} = \quad Sett_k$

  $\cup \{ [loc(H_n, V)] \mid \exists\, [loc(H, U)] \in Sett_k \text{ and}$

  $\quad loc(H, U) \Rightarrow_{[u,p]} H' \Rightarrow_{[v_1, p_1]} H_1 \Rightarrow \cdots \Rightarrow_{[v_n, p_n]} H_n,$

  $\quad$ where $u \in U$, $p = (lab_H(u) ::= (R, C)) \in copy(P)$,

  $\quad U' = (U \smallsetminus \{u\}) \cup V_R$ and $V \subseteq U'$ is a setting centre,

  $\quad (v_1, \ldots, v_n)$ are nodes in $U' \smallsetminus V$, in the derived order,

  $\quad$ for some $n \in \mathbb{N}$, and for all $i \in [n]$: $X_i = lab_{H'}(v_i)$,

  $\quad p_i = (X_i ::= (R_i, C_i))$, and $(R_i, C_i) \in Red(X_i)\}$

For $k \in \mathbb{N}$ such that $Sett_k = Sett_{k+l}$ for all $l \in \mathbb{N}$, $Sett_k$ will be denoted by $Sett$.

Making use of the definitions above, an hNCE grammar can be tested for confluence with the following algorithm.

4.15 Algorithm (confluence test)
INPUT: an hNCE grammar NG.

1. Turn NG into a nonterminal-ordered grammar;

2. compute the sets $Red(X)$;

3. compute the set $Sett$;

4. *test every critical candidate in Sett whether it is NG-confluent with respect to its centre.*

OUTPUT: *NG is confluent if and only if all critical candidates are.*

4.16 Main Theorem (decidability of confluence)
Given an hNCE grammar *NG*, the confluence test determines in time at most doubly exponential in the size of *NG* whether *NG* is confluent.

Outline of the proof. In the following section, we show that the reduced hypergraphs with embedding in $Red(X)$ suffice to compute *Sett*, and that the critical candidates in *Sett* allow to decide the confluence of the input grammar. Finally, we prove the termination of the algorithm. □

4.3 Correctness and Termination

In order to verify the correctness of the confluence test, first the correspondence of each reduced hypergraph with embedding to some leftmost derivation, and vice versa, is proved. Then all settings in *Sett* are shown to be settings of the grammar, and finally *Sett* is established as containing sufficiently many critical candidates to permit the correct conclusion on the confluence of the grammar. The termination of the confluence test relies on the finiteness of the sets $Red(X)$ and *Sett*. All in all, the algorithm is doubly exponential in the size of the input grammar. (The confluence test proposed in [Kle96] considered all derivations up to a certain length instead of using reduced hypergraphs with embedding, and thus was triply exponential in the size of the grammar.)

The correspondence between the reduced hypergraphs with embedding in the $Red(X)$ and certain leftmost derivations is shown in the following theorem.

4.17 Theorem (correctness of $Red(X)$)
Let $NG = (N, T, P, S)$ be a *nonterminal-ordered hNCE grammar and $X \in N$. There are a production $(X ::= (R, C)) \in P$ and a leftmost derivation $(R, C) \Rightarrow_P^* (R', C')$ if and only if there is $k \in \mathbb{N}$ with $red(R', C') \in Red(X)_k$.*

Proof. '⇐': Let $k \in \mathbb{N}$ and $(H, C_H) \in Red(X)_k$. Show that there are $(X ::= (R, C)) \in P$ and $(R, C) \Rightarrow_P^* (R', C')$ with $(H, C_H) = red(R', C')$ by induction on k.
$Red(X)_0 = \emptyset$, so there is nothing to show.
Let $(H, C_H) \in Red(X)_{k+1}$. This means that P contains a production $X ::= (R, C)$ and there is a derivation

$$(R, C) \Rightarrow_{[v_1, p_1]} (R_1, C_1) \Rightarrow \dots \Rightarrow_{[v_n, p_n]} (R_n, C_n)$$

such that $(v_1, \ldots, v_n)$ are some nonterminal nodes of R, in the fixed order, $p_i = (lab_R(v_i) ::= (R'_i, C'_i))$ with $(R'_i, C'_i) \in Red(lab_R(v_i))_k$ for all $i \in [n]$, and $(H, C_H) = red(R_n, C_n)$. By induction hypothesis, there are a production $lab_R(v_i) ::= (\bar{R}_i, \bar{C}_i)$ in P and a leftmost derivation $(\bar{R}_i, \bar{C}_i) \Rightarrow_P^* (\tilde{R}'_i, \tilde{C}'_i)$ such that $(R'_i, C'_i) = red(\tilde{R}'_i, \tilde{C}'_i)$, for each $i \in [n]$. As $(v_1, \ldots, v_n)$ are in the fixed order, the derivation

$$(R, C) \Rightarrow_{[v_1, \bar{p}_1]} (\tilde{R}_1, \tilde{C}_1) \Rightarrow \ldots \Rightarrow_{[v_n, \bar{p}_n]} (\tilde{R}_n, \tilde{C}_n)$$

with $\bar{p}_i = (lab_R(v_i) ::= (\bar{R}'_i, \bar{C}'_i))$ for all $i \in [n]$ is leftmost, and the associativity of NG implies that there is also a leftmost derivation $(R, C) \Rightarrow_P^* (\tilde{R}_n, \tilde{C}_n)$.

To see that $red(\tilde{R}_n, \tilde{C}_n) = red(R_n, C_n)$, observe that for $i \in [n]$, (R_i, C_i) differs from $(\tilde{R}_i, \tilde{C}_i)$ in that the hyperedges of $\bar{R}'_1, \ldots, \bar{R}'_i$ with rank greater than 0 do not appear in (R_i, C_i), nodes in $(\tilde{R}_i, \tilde{C}_i)$ having been introduced as part of the same $(\bar{R}'_j, \bar{C}'_j)$ (for some $j \in [i]$) and carrying the same label are already glued together in (R_i, C_i), and (R_i, C_i) can contain a few more isolated nodes which do not appear in any connection instruction because there may be some $(\bar{R}'_j, \bar{C}'_j)$ (with $j \in [i]$) and a label in $N \cup T$ which is not assigned to any node in $\bar{R}'_j$. These differences are obliterated by the construction of the reduced hypergraphs, so that $red(\tilde{R}_n, \tilde{C}_n) = red(R_n, C_n)$.

'$\Rightarrow$': Let $(X ::= (R, C)) \in P$ and $(R, C) \Rightarrow_P^n (R', C')$ a leftmost derivation. Show that $red(R', C') \in Red(X)_k$ for some $k \in \mathbb{N}$ by induction on n.

If $n = 0$, then $red(R', C') = red(R, C) \in Red(X)_1$.

If $n > 0$, there are m nodes $v_1, \ldots, v_m$ of R rewritten in this derivation, in that order, for some $m \in \mathbb{N}$. Then the derivation consists of m leftmost subderivations of the form

$$(R_{i-1}, C_{i-1}) \Rightarrow_{[v_i, p_i]} (R'_i, C'_i) \Rightarrow^{n_i} (R_i, C_i)$$

with $(R_0, C_0) = (R, C)$, $(R_n, C_n) = (R', C')$, $p_i = (X_i ::= (\bar{R}_i, \bar{C}_i)) \in P$, and $i \in [m]$. The derivation steps in $(R'_i, C'_i) \Rightarrow^{n_i} (R_i, C_i)$ rewrite only (descendants of) nodes in $\bar{R}_i$, so that the same substitutions can be used to form a leftmost derivation $(\bar{R}_i, \bar{C}_i) \Rightarrow^{n_i} (\bar{R}'_i, \bar{C}'_i)$. By associativity, this means $(R_{i-1}, C_{i-1}) \Rightarrow_{[v_i, \bar{p}_i]} (R_i, C_i)$ with $\bar{p}_i = (X_i ::= (\bar{R}'_i, \bar{C}'_i))$. As $n_i < n$, the induction hypothesis implies that for all $i \in [m]$, $red(\bar{R}'_i, \bar{C}'_i) \in Red(X_i)_k$ for some $k \in \mathbb{N}$. But then by construction of $Red(X)_{k+1}$, we have that $red(R', C') \in Red(X)_{k+1}$.   □

Before verifying that each setting in *Sett* is a setting of NG, it has to be checked that the reduction of a hypergraph with embedding preserves sufficient information for our purposes. More precisely, the following corollary states that, if a hypergraph with embedding is substituted for a centre node of a locale and none of its nodes will be used as centre node of a future locale, then its reduced version can be used instead.

4.18 Corollary (correctness of reduced hypergraphs)
Let $NG = (N, T, P, S)$ be an hNCE grammar, H_1 and H_2 hypergraphs over $N \cup T$, and $V \subseteq V_{H_1} \cap V_{H_2}$ such that $loc(H_1, V) = loc(H_2, V)$. Moreover, let v be a node in V, V^- denote $V \smallsetminus \{v\}$, and (R, C) be a hypergraph with embedding over $N \cup T$; (R, C) and $red(R, C)$ are assumed to be disjoint from H_1 and H_2. Then the following equalities hold.

(1) $loc(H_1[v/(R, C)], V^-) = loc(H_2[v/(R, C)], V^-)$

(2) $loc(H_1[v/(R, C)], V^-) = loc(H_1[v/red(R, C)], V^-)$

(3) $loc(H_1[v/(R, C)], V^-) = loc(H_2[v/red(R, C)], V^-)$

Proof. (1) Consider the situation depicted in Figure 4.6. By assumption, we have $loc(H_1, V) = L = loc(H_2, V)$. This implies by Lemma 4.5 that we also have $loc(H_1[v/(R, C)], V^-) = L' = loc(H_2[v/(R, C)], V^-)$.

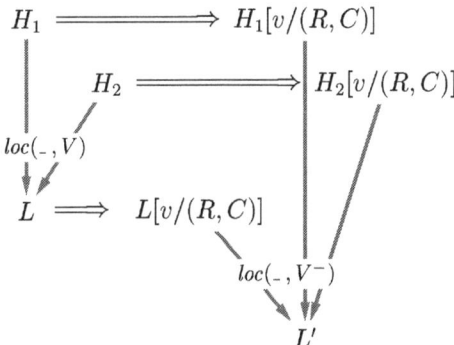

Figure 4.6: Substitution in two hypergraphs with an identical locale

(2) The two locales have the same nodes, the same hyperedges from H_1, and the same hyperedges of rank 0. There is an embedding hyperedge in $H_1[v/(R, C)]$ if and only if there is the same embedding hyperedge in $H_1[v/red(R, C)]$ which is incident to the label of a node in R instead of the node itself. As the construction of the locale implies that the nodes of (R, C) resp. $red(R, C)$ are glued with their respective labels, this difference is removed and the two locales are identical.

(3) This is a consequence of (1) and (2).   □

Now it can be shown that the construction of *Sett* is correct.

4.19 Lemma (correctness of *Sett*)
Let NG be a nonterminal-ordered hNCE grammar. Then $Sett_k \subseteq Sett(NG)$ for all $k \in \mathbb{N}$.

Proof. Show the assertion by induction on k.

Clearly, $Sett_0 = \{[loc(\bullet S, \{x\})]\} \subseteq Sett(NG)$.

Now let $[loc(H, U)] \in Sett_k$, where by induction hypothesis $H \in S(NG)$ and $U \subseteq V_H$ is a setting centre, i.e. $[loc(H, U)] \in Sett(NG)$. Consider a derivation

$$loc(H, U) \Rightarrow_{[u,p]} H_0 \Rightarrow_{[v_1,p_1]} H_1 \Rightarrow \cdots \Rightarrow_{[v_n,p_n]} H_n$$

with $u \in U$, $p = (lab_H(u) ::= (R, C)) \in P$, $U' = (U \smallsetminus \{u\}) \cup V_R$ and $V \subseteq U'$ a setting centre, $(v_1, \ldots, v_n)$ nonterminal nodes in $U' \smallsetminus V$, in the derived order, and for all $i \in [n]$: $X_i = lab_{H'}(v_i)$, $p_i = (X_i ::= (R_i, C_i))$, and $(R_i, C_i) \in Red(X_i)$. Let $H \Rightarrow_{[u,p]} H'_0$, which by Lemma 4.5 implies $loc(H_0, U') = loc(H'_0, U')$. By Theorem 4.17, there are, for all $i \in [n]$, a production $\bar{p}_i = (X_i ::= (\bar{R}_i, \bar{C}_i))$ and a derivation $(\bar{R}_i, \bar{C}_i) \Rightarrow_P^* (\bar{R}'_i, \bar{C}'_i)$ such that $red(\bar{R}'_i, \bar{C}'_i) = (R_i, C_i)$. By successive application of Corollary 4.18(3), we have for the derivation

$$H \Rightarrow_{[u,p]} H'_0 \Rightarrow_{[v_1,\bar{p}'_1]} H'_1 \Rightarrow \cdots \Rightarrow_{[v_n,\bar{p}'_n]} H'_n$$

with $\bar{p}'_i = (X_i ::= (\bar{R}'_i, \bar{C}'_i))$ that $loc(H'_i, U_i) = loc(H_i, U_i)$ for all $i \in [n]$, where $U_i = U' \smallsetminus \{v_1, \ldots, v_i\}$. Moreover, $loc(H'_n, U_n) = loc(H_n, U_n)$ means that $loc(H'_n, V) = loc(H_n, V)$ because of $V \subseteq U_n$. By associativity, each of the n derivation steps $H'_{i-1} \Rightarrow_{[v_i,\bar{p}'_i]} H'_i$ corresponds to a derivation $H'_{i-1} \Rightarrow_P^* H'_i$, so that there is a derivation $H \Rightarrow_P^* H'_n$. This implies that $[loc(H'_n, V)] = [loc(H_n, V)] \in Sett(NG)$. $\square$

The next theorem shows which settings of NG are at least in $Sett$, providing the last argument to prove the correctness of the confluence test.

4.20 Theorem ('confluence closure' of *Sett*)

Let $NG = (N, T, P, S)$ be a nonterminal-ordered hNCE grammar and let $S(NG)_n$ contain the sentential forms of NG derivable from $\bullet S$ in up to n steps. If all hypergraphs $H \in S(NG)_{n-1}$ are NG-confluent, then every setting $[loc(H, V)]$ with $H \in S(NG)_n$ is in $Sett$.

Proof. Let all sentential forms in $S(NG)_{n-1}$ be NG-confluent, and let

$$\bullet S \Rightarrow_{[v_1,p_1]} H_1 \Rightarrow \cdots \Rightarrow_{[v_n,p_n]} H_n = H$$

be a derivation of a sentential form H with $V \subseteq V_H$ a setting centre. Show that $[loc(H, V)] \in Sett$ by induction on n.

For $n = 0$, we have $S(NG)_{-1} = \emptyset$, $H = \bullet S$, $V = \{x\}$, and $[\bullet S, \{x\}] \in Sett_0$ by definition.

For $n > 0$, there are three possible cases (which are not exclusive), depending on the nature of V:

(1) V consists of a single nonterminal node $v \in V_H$: Let $l \in [n]$ be minimal such that v is a node in H_l, which means that v is generated in the derivation step

$H_{l-1} \Rightarrow_{[v_l,p_l]} H_l$ as a node in the right-hand side of $p_l = (X_l ::= (R,C))$. Let $U = \{v_l\}$.

(2) V consists of the incident nodes $vset_H(e)$ of a hyperedge $e \in E_H$ generated as part of the right-hand side of a production: Let $l \in [n]$ such that $e \in E_R$ for $p_l = (X_l ::= (R,C))$, which means that e together with its incident nodes is generated in the derivation step $H_{l-1} \Rightarrow_{[v_l,p_l]} H_l$. Let $U = \{v_l\}$.

(3) V consists of the incident nodes $vset_H(e)$ of an embedding hyperedge $e \in E_H$: Let $l \in [n]$ be maximal such that, with $p_l = (X_l ::= (R,C))$, e is created through a connection instruction in C from a hyperedge e' in H_{l-1}, which means that v_l is incident to e' in H_{l-1}. Let $U = vset_{H'_{m-1}}(e')$.

The NG-confluence of all sentential forms in $S(NG)_{n-1}$ allows to reorder the substitutions (without changing the result of the derivation) such that all substitutions $[v_i/p_i]$ with $v_i \notin U$ a node in H_{l-1} are shifted before the substitution $[v_l/p_l]$ and the derivation following this substitution is leftmost, i.e. there is a derivation

$$\bullet S \Rightarrow_P^* H'_{m-1} \Rightarrow_{[v_l,p_l]} H'_m \Rightarrow_P^* H'_n = H$$

for some $m \in \{l, \ldots, n\}$. The derivation $H'_m \Rightarrow_P^* H'_n$ consists of j leftmost sub-derivations

$$\bar{H}_{i-1} \Rightarrow_{[u_i,\bar{p}_i]} \bar{H}'_i \Rightarrow^* \bar{H}_i$$

starting with rewriting a node $u_i \in ((U \smallsetminus \{v_l\}) \cup V_R) \smallsetminus V$ with a production $\bar{p}_i = (X_i ::= (R_i, C_i))$ and rewriting only descendants of u_i ($i \in [j]$). By associativity, the derivation steps in $\bar{H}'_i \Rightarrow^* \bar{H}_i$ can be performed directly on (R_i, C_i) yielding a hypergraph with embedding (R'_i, C'_i) such that $\bar{H}_{i-1} \Rightarrow_{[u_i,\tilde{p}_i]} \bar{H}_i$ with $\tilde{p}_i = (X_i ::= (R'_i, C'_i))$. By Theorem 4.17, $red(R'_i, C'_i) \in Red(X_i)$.

Now consider the setting $[loc(H'_{m-1}, U)]$, which by induction hypothesis is in $Sett_k$ for some $k \in \mathbb{N}$, and the derivation

$$loc(H'_{m-1}, U) \Rightarrow_{[v_l,p_l]} L_0 \Rightarrow_{[u_1,\tilde{p}'_1]} L_1 \Rightarrow \ldots \Rightarrow_{[u_j,\tilde{p}'_j]} L_j,$$

where $\tilde{p}'_i = (X_i :: = red(R'_i, C'_i))$. Then we have $[loc(H,V)] = [loc(L_j, V)]$ by Corollary 4.18(3), and by construction of $Sett_{k+1}$, $[loc(L_j, V)] \in Sett_{k+1} \subseteq Sett$. □

4.21 Corollary (correctness of the confluence test)
Let NG be a nonterminal-ordered hNCE grammar.

(1) If NG is confluent, then $Sett = Sett(NG)$ and all critical candidates in $Sett$ are NG-confluent with respect to their centres.

(2) If NG is not confluent, then $Sett$ contains a critical candidate which is not NG-confluent with respect to its centre.

Proof. (1) By Lemma 4.19, $Sett \subseteq Sett(NG)$. The confluence of NG implies that all sentential forms are NG-confluent, so by Theorem 4.20 we also have $Sett(NG) \subseteq Sett$. By Theorem 4.6(2), the critical candidates in $Sett$ are NG-confluent with respect to their centres.

(2) If NG is not confluent, then there is $n \in \mathbb{N}$ minimal such that $\bullet S \Rightarrow^n_P H$ and H is not NG-confluent. This implies that $[loc(H, vset_H(e))]$ is not NG-confluent with respect to $vset_H(e)$ for some $e \in E_H$ (Theorem 4.6). Due to the minimality of n, all sentential forms in $S(NG)_{n-1}$ are NG-confluent. Thus by Theorem 4.20, $[loc(H, vset_H(e))] \in Sett$. □

Clearly, the termination of the confluence test depends on the termination of the algorithms computing the sets $Red(X)$ and $Sett$, as both transforming an hNCE grammar into a nonterminal-ordered one and testing a critical candidate for NG-confluence with respect to its centre can be done in linear time. Lemma 4.22 states that the sets $Red(X)$ can be computed in time $O(2^{(\lambda+\kappa)^\kappa})$, where κ is the maximal rank of a hyperedge in a sentential form and λ is the number of (nonterminal or terminal) symbols. Corollary 4.23 states the same for the set $Sett$.

4.22 Lemma
Let $NG = (N, T, P, S)$ be an hNCE grammar. Moreover, let $\kappa = rank(S(NG))$ and $\lambda = \#(N \cup T)$.

(1) There are less than
$$\varrho = 2^\lambda \cdot 2^{(\lambda+\kappa)^{2\kappa+4}}$$
reduced hypergraphs with embedding over $N \cup T$ such that the connection instructions describe hyperedges of rank up to κ.

(2) For all $X \in N$, $Red(X) = Red(X)_{\lambda \cdot \varrho}$.

Proof. (1) All the reduced hypergraphs have $N \cup T$ as node set. Each hyperedge is a label in $N \cup T$, of which there are λ distinct ones. Each connection instruction is an element of

$$M = \bigcup_{i=1}^{\kappa} (N \cup T) \times (N \cup T \cup \{\Diamond\})^i \times \bigcup_{j=0}^{\kappa} (N \cup T) \times (N \cup T \cup [\kappa])^j,$$

of which there are at most

$$\kappa \cdot \lambda \cdot (\lambda+1)^\kappa \cdot (\kappa+1) \cdot \lambda \cdot (\lambda+\kappa)^\kappa \; < \; (\lambda+\kappa)^{2\kappa+4}$$

distinct ones. There are 2^λ subsets of $N \cup T$ and less than $2^{(\lambda+\kappa)^{2\kappa+4}}$ subsets of M, which accounts for ϱ.

(2) By definition of the $Red(X)_k$, if there is $m \in \mathbb{N}$ with $Red(X)_m = Red(X)_{m+1}$ for all $X \in N$, then $Red(X)_m = Red(X)_{m+n}$ for all $X \in N$ and $n \in \mathbb{N}$, i.e. the

(up to λ) sets $Red(X)_i$ can only grow as long as in every iteration at least one of them grows. As the connection instructions of NG can be assumed to specify only hyperedges of rank up to κ, and as for each $X \in N$, the size of $Red(X)$ is less than ϱ by (1), this growth stops in the $\lambda \cdot \varrho$th iteration. $\qquad\square$

4.23 Corollary (termination of the confluence test)
Let $NG = (N, T, P, S)$ be a nonterminal-ordered hNCE grammar. Moreover, let $\kappa = rank(S(NG)) \geq 1$ and $\lambda = \#(N \cup T)$.
  For all $i \in \mathbb{N}$, $Sett_{\sigma+i} = Sett_\sigma$, where $\sigma = \kappa \cdot \lambda^\kappa \cdot 2^{(\kappa+\lambda)^{3\kappa}}$.

Proof. The definition of the $Sett_k$ implies that if there is $m \in \mathbb{N}$ with $Sett_m = Sett_{m+1}$, then $Sett_m = Sett_{m+n}$ for all $n \in \mathbb{N}$. As by Lemma 4.19 $Sett_k \subseteq Sett(NG)$ for all $k \in \mathbb{N}$ and by Lemma 4.9 the size of $Sett(NG)$ is bounded by σ as above, we have $Sett_{\sigma+1} = Sett_\sigma$. $\qquad\square$

4.4   Concluding Remarks

Confluence, and hence context-freeness, is decidable for edNCE grammars [Kau85] as well as for hNCE grammars [Kle96]. However, the high flexibility offered by hNCE rewriting to specify embedding hyperedges requires an algorithm which is significantly more complex than that for edNCE grammars: it runs in time doubly exponential in the size of the input grammar.

In order to get a tractable notion of confluence, Engelfriet and Rozenberg define a static criterion in [ER97] (see also [JKRE82, Definition 5.1] and [Cou87, Lemma 3.11]): An edNCE grammar $NG = (N, T, P, S)$ is *statically confluent* if for all productions $X_1 ::= (R_1, C_1)$, $X_2 ::= (R_2, C_2)$ in $copy(P)$, all nodes $v_1 \in V_1$, $v_2 \in V_2$ and all labels $\alpha, \delta \in \Sigma$, the following equivalence holds:

$$\exists \beta \in \Sigma \colon (\alpha, X_2, d/\beta, v_1, d') \in C_1 \wedge (\beta, lab_1(v_1), \overline{d'}/\delta, v_2, \overline{d'''}) \in C_2$$

$$\Longleftrightarrow$$

$$\exists \gamma \in \Sigma \colon (\alpha, X_1, \overline{d}/\gamma, v_2, \overline{d''}) \in C_2 \wedge (\gamma, lab_2(v_2), d''/\delta, v_1, d''') \in C_1$$

where $\overline{d}$ denotes the direction inverse to $d \in \{in, out\}$. This situation is sketched in Figure 4.7, where the substitution of (R_i, C_i) for the X_i-labelled nodes makes the top edge evolve to the bottom edge via either of the middle edges (the directions d, d', d'', d''' are to be read from left to right).

Clearly, every edNCE grammar which satisfies this criterion is also confluent as defined in Section 2.4. The converse does not hold, but there is a statically confluent normal form for every confluent edNCE grammar, see [ER97, Proposition 1.3.6]: For every edNCE grammar NG (confluent or not) a statically confluent edNCE

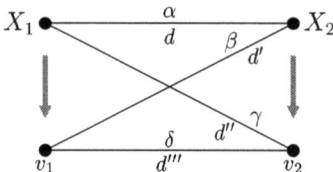

Figure 4.7: The criterion for static confluence

grammar NG' can be constructed with $L(NG') = L_{lm}(NG)$, where $L_{lm}(NG)$ denotes the language generated by leftmost derivations only (for an arbitrary but fixed order of the nodes in the right-hand sides of the productions in NG). As $L(NG) = L_{lm}(NG)$ for all C-edNCE grammars NG, this is already the desired normal form.

A similar notion of static confluence can be defined for form-preserving hNCE grammars, where a hyperedge is transformed into an embedding hyperedge by handing the tentacles gripping to the replaced node over to nodes of the replacing hypergraph and possibly changing the label; as examples consider the hNCE grammars constructed from S-HH grammars as in the proof of Theorem 3.29. It seems that link-preserving C-hNCE grammars (and thus also remote-free C-hNCE grammars) can be transformed into a form-preserving normal form, so a static definition for confluence makes sense for remote-free hNCE grammars. Still, it is unclear whether this also works for hNCE grammars containing remote connection instructions.

5

A Combination of Node and Hyperedge Rewriting

The two most prominent approaches to context-free (hyper)graph rewriting are confluent node rewriting and hyperedge rewriting, see Engelfriet [Eng97]. Each has particular strengths and weaknesses: Whereas in a node rewriting step incident hyperedges can be multiplied with some constant factor, hyperedge rewriting only allows for a constant number of additional hyperedges. On the other hand, during the rewriting of a node its incident hyperedges are grouped into a finite number of equivalence classes and cannot be distinguished beyond membership to different classes, whereas in hyperedge rewriting each hyperedge can be treated individually. In this chapter, atom replacement [HK98], a combination of these two approaches, is studied. A second combining approach has been proposed in [Bar99, Section 2.2.2].

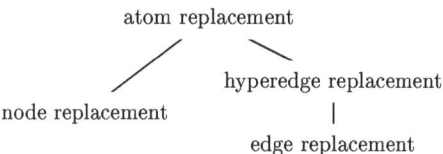

A combination may be done just by allowing both node- and hyperedge-rewriting rules within the rewriting process. Oriented by Kreowski and Rozenberg's structured graph grammars [KR90], the concept of atom replacement goes one step further by integrating connecting and gluing aspects, i.e. hypergraphs are equipped with a connection relation as well as a gluing relation, and replacing an atom in a hypergraph, i.e. a node or a hyperedge, with a hypergraph means (1) removing the atom, (2) adding the replacing hypergraph, (3) connecting it with the remainder by adding some edges according to a connection relation, and (4) gluing certain nodes of the replacing hypergraph and the remainder according to a given gluing relation. The concept subsumes node rewriting and hyperedge rewriting, and the class of generated languages, which contains a number of important sets of (hyper)graphs, has good closure properties with respect to hyperedge substitution.

In Section 5.1, atom replacement is introduced. The relationship to hNCE rewriting and HR rewriting is studied in Section 5.2; in particular, it turns out that the possibility to insert embedding hyperedges does not augment the generative power of hyperedge rewriting. Section 5.3 contains a result stating that atom-replacement languages are closed under hyperedge substitution with certain types of hypergraph languages. This yields a method to construct several important sets of graphs resp. hypergraphs with associative atom-replacement grammars. Finally, Section 5.4 contains some concluding remarks.

5.1   Atom Replacement

Atom replacement is a combination of node rewriting and hyperedge rewriting where an atom—a node or a hyperedge—can be replaced with a hypergraph [HK98]. The approach allows to glue as well as connect when rewriting a node, as sketched in Figure 5.1, and to connect as well as glue when rewriting a hyperedge, as sketched in Figure 5.2. In these figures, the white atom is the replaced one, and the gluing of two nodes is represented by a broad grey line linking these nodes.

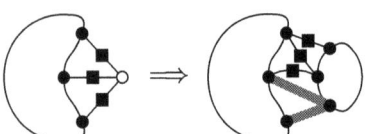   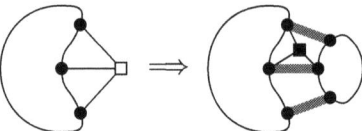

Figure 5.1: Atom replacement: rewriting a node

Figure 5.2: Atom replacement: rewriting a hyperedge

The connecting resp. gluing is controlled by specifying nodes through hyperedges: Whether a hyperedge is incident to the replaced node or is replaced itself, its incident nodes in the context of the replaced atom can be made incident to an embedding hyperedge as in hNCE rewriting, and also glued with nodes of the replacing hypergraph as in HR rewriting.

Formally, we consider hypergraphs which are extended by both a connection relation and a gluing relation.

5.1 Definition (hypergraph with embedding and gluing)

Let Σ be a finite set of symbols. A *hypergraph with embedding and gluing* over Σ is a triple $(H, C, glue)$ where H is a multiple hypergraph over Σ, the connection relation $C \subseteq CI_H$ is a finite set of connection instructions, and $glue \subseteq \bigcup_{i \in \mathbb{N}} \Sigma^i \times V_H{}^i$ is a finite *gluing relation* with, for each *gluing instruction* $(a_1 \ldots a_k / v_1 \ldots v_k) \in glue$, $a_1 \ldots a_k = lab_H(v_1 \ldots v_k)$. The set of all hypergraphs with embedding and gluing over Σ is denoted by $\mathcal{MEG}_\Sigma$.

Two hypergraphs with embedding and gluing $(H_1, C_1, glue_1)$ and $(H_2, C_2, glue_2)$ over Σ are isomorphic if there is a hypergraph isomorphism $f \colon H_1 \to H_2$ such that

$$C_2 = \{(ex/cr_2) \mid (ex/cr_1) \in C_1,\ lab(cr_2) = lab(cr_1),\ rank(cr_2) = rank(cr_1),$$
$$\forall i \in [rank(cr_2)] \colon cr_2[i] = f(cr_1[i])\ \text{if}\ cr_1[i] \in V_{H_1}\ \text{and}$$
$$cr_2[i] = cr_1[i] \quad \text{if}\ cr_1[i] \in \mathbb{N}_+\}$$

and $glue_2 = \{(a_1 \ldots a_k / f^*(v_1 \ldots v_k)) \mid (a_1 \ldots a_k / v_1 \ldots v_k) \in glue_1,\ k \in \mathbb{N}\}$.

As usual, a hypergraph with embedding and gluing $(H, \emptyset, \emptyset)$ will be considered the same as the multiple hypergraph H, so that in this sense $\mathcal{M}_\Sigma \subseteq \mathcal{MEG}_\Sigma$.

5.2 Example (hypergraph with embedding and gluing)

If H is a multiple hypergraph which contains (at least) an a-labelled node u_1 and two b-labelled nodes u_2, u_3, then $(H, C, glue)$ with

$$C = \{(c, a\Diamond/c, 1u_1), (c, a\Diamond/c, u_21), (c, \Diamond b/d, u_22), (X, ab\Diamond b/X, 1u_22)\}$$

and

$$glue = \{(b/u_3), (abb/u_1u_2u_3)\}$$

is a hypergraph with embedding and gluing. It is used in Figures 5.3 and 5.4 to illustrate atom replacement, replacing a node resp. a hyperedge.   ∎

In order to define the replacement of an atom, the following two mappings on strings will be used.

5.3 Definition

Let A^* contain the finite sequences over a set A. The mapping $-_a \colon A^* \to A^*$ erases all occurrences of some $a \in A$, i.e. it is the string homomorphism induced by $-_a (a) = \lambda$ and $-_a (b) = b$ for $b \in A \smallsetminus \{a\}$. The mapping $\#_a \colon A^* \to \mathbb{N}$ counts the occurrences of some $a \in A$, i.e. it is recursively defined by $\#_a(\lambda) = 0$, $\#_a(aw) = \#_a(w) + 1$, and $\#_a(bw) = \#_a(w)$ for all $w \in A^*$ and $b \in A \smallsetminus \{a\}$.

Replacing a node with a hypergraph with embedding and gluing means replacing the node as in the hNCE approach (but with multiple hypergraphs), *plus* gluing the hypergraph into the remainder hypergraph. Figure 5.3, a more detailed version of Figure 5.1, shows such a replacement.

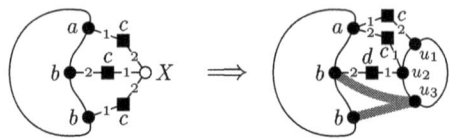

Figure 5.3: Replacement of a node

5.4 Definition (replacement of a node)

Let H_1 be a multiple hypergraph containing a node x and $(H_2, C_2, glue_2)$ a hypergraph with embedding and gluing disjoint from H_1. Then the multiple hypergraph $H_3 = H_1[x/(H_2, C_2, glue_2)]$ is obtained as a result of the following basic steps:

1. REMOVE x together with all incident hyperedges from H_1, yielding the remainder H_1^- of H_1.

2. ADD H_2 to H_1^-.

3. CONNECT H_2 and H_1^- with hyperedges according to the connection relation C_2: for all hyperedges $e \in E_{H_1}$ incident with x and all connection instructions $(ex/cr) \in C_2$ such that

   - $lab_{H_1}(e) = lab(ex)$,
   - $rank_{H_1}(e) = rank(ex)$, and
   - for all $i \in [rank(ex)]$, $ex[i] = \Diamond$ if and only if $att_{H_1}(e, i) = x$, and $ex[i] = lab_{H_1}(att_{H_1}(e, i))$ otherwise,

   add a hyperedge e' with

   - $lab_{H_3}(e') = lab(cr)$,
   - $rank_{H_3}(e') = rank(cr)$, and
   - for all $j \in [rank(cr)]$, $att_{H_3}(e', j) = cr[j]$ if $cr[j] \in V_{H_2}$ and $att_{H_3}(e', j) = att_{H_1}(e, cr[j])$ if $cr[j] \in \mathbb{N}_+$.

4. GLUE H_1^- and H_2 according to the gluing relation $glue_2$: for all hyperedges $e \in E_{H_1}$ incident with x and all gluing instructions $(a_1 \ldots a_k/u_1 \ldots u_k) \in glue_2$ with $lab_{H_1}^*(-_x(att_{H_1}(e))) = a_1 \ldots a_k$, identify, for all $i \in [k]$, the ith node in $-_x(att_{H_1}(e))$ and u_i.

Replacing a hyperedge with a hypergraph with embedding and gluing means replacing the hyperedge as in the HR approach (but with node labels), *plus* embedding the hypergraph into the remainder hypergraph. Figure 5.4, a more detailed version of Figure 5.2, shows such a replacement.

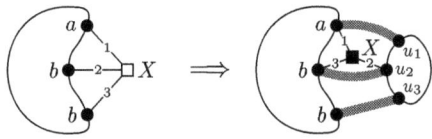

Figure 5.4: Replacement of a hyperedge

5.5 Definition (replacement of a hyperedge)

Let H_1 be a multiple hypergraph containing a hyperedge x and $(H_2, C_2, glue_2)$ a hypergraph with embedding and gluing disjoint from H_1. Then the multiple hypergraph $H_3 = H_1[x/(H_2, C_2, glue_2)]$ is obtained as a result of the following basic steps:

1. REMOVE the hyperedge x from H_1, yielding the remainder H_1^- of H_1.

2. ADD H_2 to H_1^-.

3. GLUE H_1^- and H_2 according to the gluing relation $glue_2$: for all gluing instructions $(a_1 \ldots a_k / u_1 \ldots u_k) \in glue_2$ with $lab_{H_1}^*(att_{H_1}(x)) = a_1 \ldots a_k$, identify, for all $i \in [k]$, $att_{H_1}(x, i)$ and u_i.

4. CONNECT H_2 and H_1^- with hyperedges according to the connection relation C_2: for each connection instruction $(ex/cr) \in C_2$ with

   - $lab_{H_1}(x) = lab(ex)$ and
   - $lab_{H_1}^*(att_{H_1}(x)) = -_\diamond (ex[1..rank(ex)])$,

   add a hyperedge e' with

   - $lab_{H_3}(e') = lab(cr)$,
   - $rank_{H_3}(e') = rank(cr)$, and
   - for all $i \in [rank(cr)]$, $att_{H_3}(e', i) = cr[i]$ if $cr[i] \in V_{H_2}$, and $att_{H_3}(e', i) = att_{H_1}(x, k)$ (where $k = cr[i] - \#_\diamond(x_1 \ldots x_{ex[i]})$) if $cr[i] \in \mathbb{N}_+$.

Note that independently of whether the replaced atom is a node or a hyperedge, the results of the embedding and the gluing steps depend solely on the immediate context of the atom. Thus, these steps do not interfere with each other and can be performed in any order.

As atom replacement allows to rewrite nodes as well as hyperedges, there is no standard form for the axiom as in the case of hNCE, HR, or HH grammars. Therefore, the axiom may now be any multiple hypergraph.

5.6 Definition (atom-replacement grammar)

An *atom-replacement grammar* (AR grammar for short) is a tuple $AG = (N, T, P, Z)$ where N and T are finite, disjoint sets of nonterminal and terminal symbols respectively, $P \subseteq N \times \mathcal{MEG}_{N \cup T}$ is a finite set of productions, and $Z \in \mathcal{M}_{N \cup T}$ is the *axiom*. The class of hypergraph languages generated by atom-replacement grammars is denoted by $\mathcal{L}(\mathrm{AR})$.

Derivations, sentential forms, and generated multiple hypergraphs are defined as usual.

5.7 Example (AR grammars)

1. The set COMPLETE of all complete graphs can be generated by the AR grammar AG_1 with a unique nonterminal label S, a terminal label a, $\bullet S$ as the axiom, and productions

$$S ::=_{p_1} (\, a \bullet \underset{u_1 \;\; a \;\; u_2}{\overset{a}{\rightleftarrows}} \bullet S \;, C_1, \emptyset) \qquad \text{and} \qquad S ::=_{p_2} (\, \bullet \underset{u}{a} \;, C_2, \emptyset)$$

with C_1 and C_2 as in Example 2.9.1. The derivation shown in Figure 2.10 is also a derivation in AG_1.

2. For $k \in \mathbb{N}_+$, the set PARTkTREE of partial k-trees can be generated by an AR grammar. For $k = 2$, such a grammar is AG_2 with a unique nonterminal label X, a unique terminal label $*$, production

$$X ::=_{p_1} (\, \overset{u_2 \bullet \overset{X}{} \bullet u_1}{\underset{X \diagdown X}{\bullet}} \;, \emptyset, glue)$$

with $glue = \{(**/u_1 u_2)\}$ to build a directed 2-tree with nonterminal edges, productions

$$X ::=_{p_2} (\, \underset{u_1}{\bullet} \longrightarrow \underset{u_2}{\bullet} \;, \emptyset, glue) \qquad X ::=_{p_4} (\, \underset{u_1}{\bullet} \quad \underset{u_2}{\bullet} \;, \emptyset, glue)$$

$$X ::=_{p_3} (\, \underset{u_1}{\bullet} \longleftarrow \underset{u_2}{\bullet} \;, \emptyset, glue) \qquad X ::=_{p_5} (\, \underset{u_1}{\bullet} \rightleftarrows \underset{u_2}{\bullet} \;, \emptyset, glue)$$

to terminate the edges, and the graph in the right-hand side of p_1 as axiom. A derivation in AG_2 is shown in Figure 5.5.

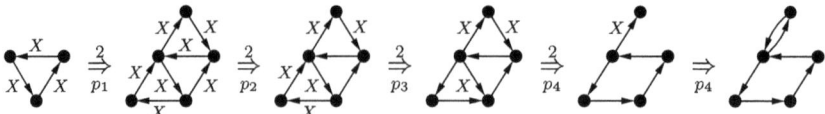

Figure 5.5: A derivation in AG_2 yielding a partial 2-tree

3. The set $plus(\text{COMPLETE})$ obtained from the complete graphs by adding a third incident node to every edge can be generated by the AR grammar AG_3 with a nonterminal node label S, a nonterminal edge label X, a unique terminal label $*$, the axiom $\bullet S$, productions analogous to those of AG_1 to build complete graphs with unlabelled nodes and X-labelled edges, i.e.

$$S ::= \underset{p_1}{(\overset{u_1}{\bullet} \overset{}{\longrightarrow} \overset{u_2}{\bullet} S, C_1, \emptyset)} \quad \text{and} \quad S ::= \underset{p_2}{(\overset{}{\bullet}_u, C_2, \emptyset)}$$

with $C_1 = \{(ex_{in}/X, 1u_1), (ex_{out}/X, u_12), (ex_{in}/*, 1u_2), (ex_{out}/*, u_22)\}$, $C_2 = \{(ex_{in}/X, 1u), (ex_{out}/X, u2)\}$, $ex_{in} = (*, *\Diamond)$, and $ex_{out} = (*, \Diamond*)$, and a production to add the third incident node to every edge, i.e.

$$X ::= \underset{p_3}{(\overset{u_1}{\bullet}\overset{1}{\underset{u_2}{\bullet}}_2 \blacksquare\!-\!3\!-\!\bullet, \emptyset, \{(**/u_1u_2)\})} \,.$$

A derivation in AG_3 is shown in Figure 5.6. ∎

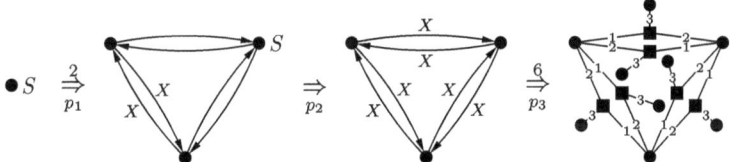

Figure 5.6: A derivation in AG_3

5.2 Relationship to the Original Concepts

Atom replacement subsumes node rewriting as well as hyperedge rewriting, i.e. an hNCE resp. HR grammar can be transformed into an AR grammar generating the same language (up to parallel hyperedges in the hNCE case). Moreover, there is a normal form for AR grammars such that, whenever in a derivation a hyperedge is rewritten, the production has an empty connection relation and a gluing relation consisting of a unique instruction. This implies that embedding hyperedges do not augment the generative power of hyperedge rewriting.

An hNCE or an HR grammar induces an (equivalent) AR grammar in a straight-forward way.

5.8 Definition (induced AR grammar)
An AR grammar $Gr' = (N, T, P', Z)$ is *induced* by a grammar $Gr = (N, T, P, S)$ if

- Gr is an hNCE grammar,[1] $P' = \{X ::= (R, C, \emptyset) \mid (X ::= (R, C)) \in P\}$, and $Z = \bullet S$; or

- Gr is an HR grammar, $P' = \{X ::= (R, \emptyset, glue) \mid (X ::= (R, ext)) \in P$, $glue = \{(*^{|ext|}/ext)\}\}$, and $Z = \blacksquare S$.

The AR grammar AG_1 from Example 5.7 is induced by the hNCE grammar NG_1 from Example 2.9.1. The grammar AG_2 becomes induced by an HR grammar if a nonterminal S and a production $S ::= Z$ are added (where Z is the axiom of AG_2), and the axiom is changed to $\blacksquare S$.

While hNCE grammars generate simple hypergraphs, AR grammars generate multiple hypergraphs. A basis to compare their generative power is provided by flattening the multiple hypergraphs of the AR languages to simple hypergraphs, i.e. by identifying parallel hyperedges in multiple hypergraphs.

5.9 Definition (flattening of multiple hypergraphs)

The mapping $flat \colon \mathcal{M}_\Sigma \to \mathcal{H}_\Sigma$ assigns to every multiple hypergraph $H = (V, E, lab, att)$ the simple hypergraph $flat(H)$ with the same nodes, labelled as in H, and the hyperedge set $\{(lab(e), att(e)) \mid e \in E\}$.

Now the expected relationship between hNCE and AR rewriting can be stated.

5.10 Lemma (simulation of hNCE rewriting)

Let NG be an hNCE grammar and NG' the induced AR grammar. Then $L(NG) = flat(L(NG'))$.

Proof. In both NG and NG', only nodes can be rewritten. This implies a one-to-one correspondence between the derivations in NG and NG'. As any two parallel hyperedges behave identically during a rewriting step in NG', it can be shown by an induction on the length of the derivations that a sentential form H of NG differs from its counterpart H' of NG' only in that any parallel hyperedges which occur in H' are identified in H, i.e. $H = flat(H')$. Of course, this also holds for terminal sentential forms. $\square$

The immediate consequence of Lemma 5.10 is that up to parallel hyperedges, hNCE languages are AR languages.

5.11 Corollary

$\mathcal{L}(hNCE) \subseteq flat(\mathcal{L}(AR))$.

---

[1]In order to keep the following construction simple, we assume in this section that for every hNCE grammar a partition of the nonterminal labels into sets N_V and N_E exists such that all nonterminal node labels occurring in the grammar, including the left-hand sides of the productions, are taken from N_V, and all nonterminal hyperedge labels from N_E. This normal form for hNCE grammars can be constructed easily, and would otherwise have to be integrated in the construction of the induced AR grammar.

It is an open question whether the inclusion is proper.

As the languages generated by hyperedge rewriting consist of multiple hypergraphs, HR grammars can be simulated directly.

5.12 Lemma (simulation of HR rewriting)
Let HG be an HR grammar and HG' the induced AR grammar. Then $L(HG) = L(HG')$.

Proof. As the unique node label $*$ is terminal, only hyperedges can be rewritten in HG'. There is a one-to-one correspondence of the derivations in HG and HG', and two corresponding derivations generate the same (terminal) hypergraph. □

5.13 Corollary
$\mathcal{L}(HR) \subsetneq \mathcal{L}(AR)$.

Proof. The inclusion is implied by Lemma 5.12. Example 5.7.1 shows that COMPLETE $\in \mathcal{L}(AR)$, but COMPLETE cannot be generated by an HR grammar, see Theorem 2.6 in [Hab92a, Chapter IV]. □

Having stated that atom replacement generalises the concepts of node and hyperedge rewriting, the next question is which part of atom replacement (if any) leads to augmented generative power: the combination of embedding and gluing in one derivation step, or the combination of node rewriting steps and hyperedge rewriting steps in one derivation? While the question whether the possibility to glue nodes gives more expressive power to node rewriting is still open, it is shown below that the possibility to introduce embedding hyperedges does not augment the generative power of hyperedge rewriting. For this, a normal form for AR grammars is first established.

5.14 Definition (hyperedge-rewriting simplified AR grammar)
An AR grammar $AG = (N, T, P, Z)$ is *hyperedge-rewriting simplified* if $N = N_V \uplus N_E$ such that only labels in N_V occur as nonterminal node labels and only labels in N_E occur as nonterminal hyperedge labels, and for each production $X ::= (R, C, glue)$ with $X \in N_E$, C is empty and $glue$ contains exactly one element.

5.15 Theorem (hyperedge-rewriting simplified normal form)
For every AR grammar, an equivalent hyperedge-rewriting simplified AR grammar can be constructed.

Proof. Let $AG = (N, T, P, Z)$ be an AR grammar. We will first construct an equivalent AR grammar AG' where every nonterminal hyperedge has the labels of its incident nodes integrated in its own label, and from AG' an equivalent AR grammar AG'' where, in every hyperedge-rewriting production, possible embedding hyperedges are integrated in the right-hand side hypergraph, and all gluing components are combined into one.

The grammar $AG' = (N', T, P', Z')$ is constructed from AG as follows. Let $N' = N \uplus N_E$ with $N_E = N \times \bigcup_{i\in[k]}(N\cup T)^i$, where k is the maximal rank of a hyperedge occurring in either a hypergraph of AG or the connection relations of its productions. Symbols in N will be used exclusively to label nodes, and symbols in N_E exclusively to label hyperedges. For a hypergraph H over $N \cup T$, let $\tau(H)$ be the hypergraph over N' obtained from H by changing the label $lab_H(e) \in N$ of each nonterminal hyperedge to $lab_{\tau(H)}(e) = (lab_H(e), lab_H^*(att_H(e))) \in N_E$. For every production $X ::= (R, C, glue)$ in P, P' contains the node-rewriting production $X ::= (\tau(R), C', glue)$ and for each $(X, a_1 \ldots a_n) \in N_E$ the hyperedge-rewriting production $(X, a_1 \ldots a_n) ::= (\tau(R), C'', glue)$. Here, C' is obtained from C by augmenting, in each connection instruction $(ex/cr) \in C$, the hyperedge label $lab(ex)$ (resp. $lab(cr)$) with the corresponding node labels if $lab(ex) \in N$ (resp. $lab(cr) \in N$); and C'' is obtained from C by deleting, in each connection instruction $(ex/cr) \in C$, the occurrences of $\Diamond$ in $ex[1..rank(ex)]$, adjusting the references in $cr[1..rank(cr)]$ accordingly, and augmenting where necessary $lab(ex)$ and $lab(cr)$ as described for C'. The gluing relation $glue$ remains unchanged because it does not contain (nonterminal) hyperedge labels. Finally, let $Z' = \tau(Z)$. It is straightforward to see that the derivations in AG' correspond one-to-one to those in AG, and in such a way that the counterpart of a sentential form $H \in S(AG)$ in $S(AG')$ is $\tau(H)$. As $\tau(H) = H$ for terminal hypergraphs H, this implies $L(AG) = L(AG')$.

The grammar $AG'' = (N', T, P'', Z')$ is now constructed from AG' as follows. Let $p = ((X, a_1 \ldots a_n) ::= (R, C, glue))$ be a production in P' with $(X, a_1 \ldots a_n) \in N_E$. As p can only be applied to a hyperedge e with $lab(e) = (X, a_1 \ldots a_n)$, i.e. with $lab^*(att(e)) = a_1 \ldots a_n$, every connection instruction $(ex/cr) \in C$ with $lab(ex) \neq (X, a_1 \ldots a_n)$ or $-_\Diamond (ex[1..rank(ex)]) \neq a_1 \ldots a_n$ cannot generate an embedding hyperedge, so that it can be removed from C. Next, add n nodes $u_1, \ldots, u_n$ with respective labels $a_1, \ldots, a_n$ disjointly to R, yielding R', and a gluing component $(a_1 \ldots a_n/u_1 \ldots u_n)$ to $glue$, yielding $glue'$; this does not change the result of applying the production. For each of the connection instructions $coin = (ex/cr)$ remaining in C, add a hyperedge e_{coin} to R' with $lab(e_{coin}) = lab(cr)$ and, for all $i \in [rank(cr)]$, $att(e_{coin}, i) = cr[i]$ if $cr[i] \in V_R$, and $att(e_{coin}, i) = u_j$ if $cr[i] \in \mathbb{N}_+$ and $j = cr[i] - \#_\Diamond(a_1 \ldots a_{cr[i]})$; and then remove $coin$ from C. Again, this does not change the result of applying the production, and the resulting production $(X, a_1 \ldots a_n) ::= (R'', C', glue')$ has an empty connection relation. Finally, consider the gluing relation. All gluing components $(a'_1 \ldots a'_k/v_1 \ldots v_k) \in glue'$ with $a'_1 \ldots a'_k \neq a_1 \ldots a_n$ will never lead to any gluing and can be deleted. For all gluing components $(a_1 \ldots a_n/v_1 \ldots v_n) \in glue' \smallsetminus \{(a_1 \ldots a_n/u_1 \ldots u_n)\}$ which are left, if $v_i \neq u_i$ for some $i \in [n]$, then glue v_i and u_i in R'' and substitute u_i for v_i in all remaining gluing components. Only $(a_1 \ldots a_n / u_1 \ldots u_n)$ is left after that, and as some of the u_i may have been glued with each other in this stage, the sequence $u_1 \ldots u_n$ has to be adjusted accordingly. This transformation does not change the result of applying the production either, and only the gluing instruction obtained from $(a_1 \ldots a_n / u_1 \ldots u_n)$ belongs to the gluing relation thus constructed.

With every hyperedge-rewriting production of AG' transformed like this, the grammar AG'' is obtained which is hyperedge-rewriting simplified and generates the same language as AG. $\square$

If, as in [Hab92b, Section 4], HR rewriting is considered for hypergraphs where nodes as well as hyperedges are labelled, the previous lemma immediately implies that every AR grammar where in every derivation only hyperedges are rewritten can be simulated by an HR grammar.

5.16 Corollary (simulation of hyperedge-rewriting AR grammars)
Let AG be an AR grammar where no node of the hypergraphs in AG is labelled with the left-hand side of a production. Then there is an HR grammar HG such that $L(HG) = L(AG)$ (up to node labelling).

Proof. Let AG be a hyperedge-rewriting AR grammar where no node of the hypergraphs in AG is labelled with the left-hand side of a production. By Theorem 5.15, AG may be assumed to be in hyperedge-rewriting simplified normal form. Then all productions are hyperedge-rewriting and have the form $(X, a_1 \ldots a_n) ::= (R, \emptyset, \{(a_1 \ldots a_n / u_1 \ldots u_n)\})$. Construct an equivalent general HR grammar (where in the right-hand side of a production, a node may occur more than once in the sequence of external nodes) by transforming every production as above into the HR production $(X, a_1 \ldots a_n) ::= (R, u_1 \ldots u_n)$, adding a new nonterminal hyperedge label S as start symbol, and adding a production $S ::= (Z, \lambda)$. By Theorem 4.6 in [Hab92a, Chapter I], this grammar can then be transformed into its so-called well-formed normal form, where the sequences of external nodes in the right-hand sides of the productions are repetition-free. $\square$

5.3 Closure Under Hyperedge Substitution

In this section, the closure of hypergraph languages under hyperedge substitution is studied. For this, two subclasses of multiple hypergraphs with embedding and gluing are introduced, namely basic multiple hypergraphs with embedding resp. with gluing. The latter can be used to define a notion of hyperedge substitution in the usual language-theoretic way. For this, a subclass of AR languages is defined so that AR languages are closed under hyperedge substitution with these languages, thereby allowing to construct several interesting (hyper)graph languages from very simple languages.

5.17 Definition (basic hypergraph with embedding resp. gluing)
Let N_V, N_E, and T be pairwise disjoint finite sets of nonterminal node, nonterminal hyperedge, and terminal labels, respectively. A hypergraph with embedding and gluing $(H, C, glue) \in \mathcal{MEG}_{N_V \cup N_E \cup T}$ is

- *nonterminal disjoint* if

    - no node is labelled in N_E, and no hyperedge in N_V,

    - a nonterminal node and a nonterminal hyperedge are not incident,

    - for all connection instructions $(ex/cr) \in C$, $lab(ex)$ is terminal, and $lab(cr)$ is in $N_E \cup T$ and nonterminal only if all nodes specified by $cr[1..rank(cr)]$ are terminal, and

    - $glue \subseteq T^* \times V_H{}^*$ allows to identify only terminal nodes;

- a *basic hypergraph with embedding* over N_V, N_E, and T if it is nonterminal disjoint and $glue$ is empty;

- a *basic hypergraph with gluing* over N_V, N_E, and T if it is nonterminal disjoint and C is empty.

The sets of basic hypergraphs with embedding resp. with gluing over N_V, N_E, and T are denoted by $\mathcal{BME}_{N_V,N_E,T}$ resp. $\mathcal{BMG}_{N_V,N_E,T}$. The set of basic hypergraphs with embedding and gluing over N_V, N_E, and T is defined as $\mathcal{BMEG}_{N_V,N_E,T} = \mathcal{BME}_{N_V,N_E,T} \cup \mathcal{BMG}_{N_V,N_E,T}$.

Given basic hypergraphs with gluing, it is possible to define a notion of hyperedge substitution; see also Section 1 in [Hab92a, Chapter III] where this notion is introduced for node-unlabelled hypergraphs.

5.18 Definition (hyperedge substitution)
Let T and $\Gamma \subseteq T$ be finite sets of symbols. A mapping $subst \colon \Gamma \to [\mathcal{BMG}_{\emptyset,\emptyset,T}]$ is called a *hyperedge substitution*. It is extended to abstract multiple hypergraphs $[H] \in [\mathcal{M}_T]$ by defining

$$subst([H]) = \{[H[e_1/(H_1, \emptyset, glue_1)] \ldots [e_n/(H_n, \emptyset, glue_n)]] \mid$$
$$e_1, \ldots, e_n \text{ are the hyperedges in } H \text{ with label in } \Gamma,$$
$$[(H_i, \emptyset, glue_i)] \in subst(lab_H(e_i)) \text{ for all } i \in [n],$$
$$H \text{ and the } H_i \text{ are pairwise disjoint}\}.$$

A hyperedge substitution $subst \colon \Gamma \to \mathcal{BMG}_{\emptyset,\emptyset,T}$ is *finite* if for all $\gamma \in \Gamma$, $subst(\gamma)$ is finite; it is a *hyperedge homomorphism* if each $subst(\gamma)$ contains exactly one element.

5.19 Example (hyperedge substitution)
Let $T = \{a, *\}$, and let H be the complete graph with three nodes (and six edges), where nodes are labelled with $*$ and edges with a. Moreover, let $subst \colon \{a\} \to [\mathcal{BMG}_{\emptyset,\emptyset,T}]$ be the hyperedge homomorphism assigning to a the set containing as unique element

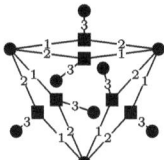

and let $subst' \colon \{a\} \to [\mathcal{BMG}_{\emptyset,\emptyset,T}]$ be the hyperedge substitution assigning to a the set containing the basic hypergraphs with gluing $[(H', \emptyset, \{(**/u_1 u_2)\})]$ where H' is a chain of edges of the form:

Then $subst([H])$ consists of the hypergraph

and $subst'([H])$ contains the following graphs, among others:

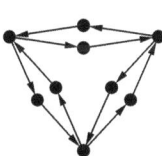 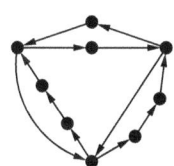 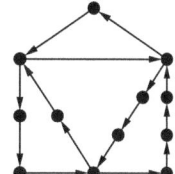

It is immediate to see that C-hNCE and C-edNCE languages are not closed under any type of (hyper)edge substitution.

5.20 Proposition (non-closure under hyperedge homomorphism)
C-hNCE and C-edNCE languages are not closed under hyperedge homomorphism.

Proof. The set of all complete a-labelled graphs can be generated by the confluent (even linear) edNCE grammar NG_1 from Example 2.9.1. Changing $subst$ from the previous example so that the hyperedge is removed from the hypergraph with which $subst$ substitutes edges, and applying the resulting hyperedge homomorphism to COMPLETE yields the set of all discrete graphs with n^2 nodes, for $n \in \mathbb{N}_+$. By Corollary 3.40, this set cannot be generated by a C-hNCE grammar, and hence neither by a C-edNCE grammar. □

As, unlike [Hab92a], the extension of hyperedge substitution is defined here for abstract hypergraphs, its well-definedness has to be verified.

5.21 Lemma (well-definedness of hyperedge substitution)
The extension of a hyperedge substitution to abstract multiple hypergraphs is well defined.

Proof. Let T and $\Gamma \subseteq T$ be finite sets of symbols and $subst \colon \Gamma \to \mathcal{BMG}_{\emptyset, \emptyset, T}$ a hyperedge substitution. Moreover, let $H \in \mathcal{M}_T$ be a multiple hypergraph with $e_1, \dots, e_n$ as hyperedges labelled in Γ, and let, for all $i \in [n]$, $[(H_i, \emptyset, glue_i)] \in subst(lab_H(e_i))$ such that H and the H_i are pairwise disjoint.

The substitutions of the $(H_i, \emptyset, glue_i)$ for the e_i are sequentially independent. Formally, if $\pi \colon [n] \to [n]$ is a permutation, then

$$H[e_1/(H_1, \emptyset, glue_1)] \dots [e_n/(H_n, \emptyset, glue_n)]$$
$$= H[e_{\pi(1)}/(H_{\pi(1)}, \emptyset, glue_{\pi(1)})] \dots [e_{\pi(n)}/(H_{\pi(n)}, \emptyset, glue_{\pi(n)})].$$

This follows from the fact that when two hyperedges are replaced with basic hypergraphs with gluing, i.e. the hyperedges are removed and their incident nodes are (possibly) glued with nodes of the respective replacing hypergraphs, then the same nodes are glued, independently of the sequence of the replacements.

If $(H_1', \emptyset, glue_1')$ is isomorphic to $(H_1, \emptyset, glue_1)$ and H_1' is disjoint with $H_2, \dots, H_n$ and H, then substituting $(H_1, \emptyset, glue_1)$ resp. $(H_1', \emptyset, glue_1')$ for e_1 yields isomorphic hypergraphs:

$$H[e_1/(H_1, \emptyset, glue_1)][e_2/(H_2, \emptyset, glue_2)] \dots [e_n/(H_n, \emptyset, glue_n)]$$
$$\cong H[e_1/(H_1', \emptyset, glue_1')][e_2/(H_2, \emptyset, glue_2)] \dots [e_n/(H_n, \emptyset, glue_n)].$$

The isomorphism which witnesses this assertion is the isomorphism between H_1 and H_1', and the identity otherwise. Together with the sequential independence of the hyperedge substitutions, we obtain that the extension of a hyperedge substitution to hypergraphs does not depend on the concrete representants of the abstract basic hypergraphs with gluing chosen to be substituted for the hyperedges.

Finally, the extension does not depend on the concrete representant chosen to get its hyperedges substituted either: If H' is isomorphic to H, disjoint from all the H_i, and e_i' is its hyperedge corresponding to e_i in H, for all $i \in [n]$, then

$$H[e_1/(H_1, \emptyset, glue_1)] \dots [e_n/(H_n, \emptyset, glue_n)]$$
$$\cong H'[e_1'/(H_1, \emptyset, glue_1)] \dots [e_n'/(H_n, \emptyset, glue_n)],$$

where the witnessing isomorphism behaves as the isomorphism between H and H' on all parts of these hypergraphs except the replaced hyperedges, and is the identity otherwise. $\qquad\square$

It is not difficult to show that AR languages are closed under finite hyperedge substitution. One would, however, like to have a more general closure, including substitutions such as *subst'* of Example 5.19 which assigns a very simple, but infinite, edge-replacement language to a. In order to prove the closure of some grammar-generated class of languages under general substitution, one usually exploits the associativity and confluence (or at least the sequential independence of certain rewriting steps) of the grammars. Unfortunately, general AR grammars have neither property: Obviously, the AR grammar induced by a non-confluent hNCE grammar is not confluent either, and Example 5.22 shows where an AR grammar may be non-associative.

5.22 Example (non-associative AR grammar)

Let AG be an AR grammar which generates as a sentential form the hypergraph $H = X \bullet\!\!-\!\!\blacksquare X$ and contains the productions $X ::= (\bullet X, \emptyset, \{(X/v)\})$ and $X ::= (\bullet a, \emptyset, \emptyset)$. The derivation starting in H where the first production is used to rewrite the hyperedge e, and the second production to rewrite the node v substituted for this hyperedge, is as follows:

$$X \bullet\!\!-\!\!\blacksquare X \Rightarrow X \bullet\!\!\!\text{-----}\!\!\bullet X = \bullet X \Rightarrow \bullet a$$

AG being associative would now mean that

$$X \bullet\!\!-\!\!\blacksquare X \, [e/\,(\bullet X, \emptyset, \{(X/v)\}) \, [v/(\bullet a, \emptyset, \emptyset)]] = \bullet a.$$

Clearly, $(\bullet X, \emptyset, \{(X/v)\})[v/(\bullet a, \emptyset, \emptyset)]$ must be defined as some hypergraph with embedding and gluing $(\bullet a, \emptyset, glue)$. As the node in this result is labelled a, there is no way to define $glue$ such that $X \bullet\!\!-\!\!\blacksquare X[e/(\bullet a, \emptyset, glue)] = \bullet a$. Hence, AG cannot be associative. ∎

Atom replacement for basic hypergraphs with embedding and gluing yields a class of associative AR grammars.

5.23 Definition (basic atom replacement)

Let N_V, N_E, and T be disjoint finite sets of symbols, and let $(H_1, C_1, glue_1) \in \mathcal{BMEG}_{N_V, N_E, T}$.

For a node $v \in V_{H_1}$ with $lab_{H_1}(v) \in N_V$ and a basic hypergraph with embedding $(H_2, C_2, \emptyset)$ with H_1 and H_2 disjoint, $(H_1, C_1, glue_1)[v/(H_2, C_2, \emptyset)]$ is the basic hypergraph with embedding and gluing $(H_3, C_3, glue_3)$ with

- $H_3 = H_1[v/(H_2, C_2, \emptyset)]$ obtained as in general atom replacement,

- C_3 as in hNCE rewriting (see Definition 2.6), and

- $glue_3 = glue_1$.

For a hyperedge $e \in E_{H_1}$ with $lab_{H_1}(e) \in N_E$ and a basic hypergraph with gluing $(H_2, \emptyset, glue_2)$ with H_1 and H_2 disjoint, $(H_1, C_1, glue_1)[e/(H_2, \emptyset, glue_2)]$ is the basic hypergraph with embedding and gluing $(H_3, C_3, glue_3)$ with

- $H_3 = H_1[e/(H_2, \emptyset, glue_2)]$ obtained as in general atom replacement,

- C_3 obtained from C_1 by updating the nodes in the creation parts of the connection instructions to the possibly glued nodes of H_3, and

- $glue_3$ obtained analogously from the gluing instructions in $glue_1$.

The following grammars, which basically combine node-rewriting and hyper-edge-rewriting steps, are associative. Moreover, the axiom may be a basic hyper-graph with gluing, so that the generated language can be used to be substituted for hyperedges.

5.24 Definition (basic atom-replacement grammar)
A *basic atom-replacement grammar* (BAR grammar for short) is a tuple $AG = (N_V \uplus N_E, T, P, (Z, \emptyset, glue_Z))$ where N_V, N_E, and T are finite, disjoint sets of nonterminal node, nonterminal hyperedge, and terminal labels respectively, $P \subseteq (N_V \times \mathcal{BME}_{N_V, N_E, T}) \cup (N_E \times \mathcal{BMG}_{N_V, N_E, T})$ is a finite set of productions, and $(Z, \emptyset, glue_Z) \in \mathcal{BMG}_{N_V, N_E, T}$ is the axiom.

5.25 Lemma (associativity of BAR grammars)
Basic atom-replacement grammars are associative.

Proof. Let $AG = (N_V \uplus N_E, T, P, (Z, \emptyset, glue_Z))$ be a BAR grammar. Moreover, let $(H, C, glue) \in \mathcal{BMEG}_{N_V, N_E, T}$, x_1 be a nonterminal atom in H, $lab_H(x_1) ::= (H_1, C_1, glue_1)$ a production (copy) in AG, x_2 a nonterminal atom in H_1, and $lab_{H_1}(x_2) ::= (H_2, C_2, glue_2)$ a production (copy) in AG such that H, H_1, and H_2 are pairwise disjoint.

If x_1 is replaced by $(H_1, C_1, glue_1)$ in $(H, C, glue)$ then every atom of H_1, and in particular x_2, belongs to the result.

In the following, let

$$(H_1', C_1', glue_1') = (H_1, C_1, glue_1)[x_2/(H_2, C_2, glue_2)],$$
$$(H', C', glue') = (H, C, glue)[x_1/(H_1', C_1', glue_1')],$$
$$(H'', C'', glue'') = (H, C, glue)[x_1/(H_1, C_1, glue_1)][x_2/(H_2, C_2, glue_2)].$$

To see that $(H', C', glue') = (H'', C'', glue'')$, four cases have to be distinguished, depending on whether x_1 resp. x_2 is a node or a hyperedge, i.e. whether $(H_1, C_1, glue_1)$ resp. $(H_2, C_2, glue_2)$ is a basic hypergraph with embedding, or with gluing.

Case 1: $x_1 \in V_H$ and $x_2 \in V_{H_1}$.
Then $glue_1 = \emptyset = glue_2$, and also $glue_1' = \emptyset$. This implies that $glue' = glue = glue''$. The proof that $H' = H''$ and $C' = C''$ can be done analogously to the hNCE case, see Lemma 2.15.

Case 2: $x_1 \in V_H$ and $x_2 \in E_{H_1}$.
Then $glue_1 = \emptyset$, $C_2 = \emptyset$, and $glue_1' = \emptyset$. This implies that $glue' = glue = glue''$. It is irrelevant whether first the hyperedges incident to x_1 in H are transformed into embedding hyperedges and then their incident nodes from H_1 are glued with nodes from H_2, or whether first these nodes are glued and then embedding hyperedges made incident to the glued nodes. Thus, $H' = H''$. Analogously, $C' = C''$.

Case 3: $x_1 \in E_H$ and $x_2 \in V_{H_1}$.
Then $C_1 = \emptyset$, $glue_2 = \emptyset$, $C_1' = \emptyset$, and $glue_1' = glue_1$. This implies that $C' = C''$ and $glue' = glue''$. As only terminal nodes can be glued by $glue_1$, x_2 must be an 'internal' node. Therefore, all its incident hyperedges and neighbours in H_1 are also present in $H[x_1/(H_1, C_1, glue_1)]$ (although some of the neighbours may be glued with other nodes), and it makes no difference whether x_2 is replaced before or after $(H_1, C_1, glue_1)$ is substituted for x_1, i.e. $H' = H''$.

Case 4: $x_1 \in E_H$ and $x_2 \in E_{H_1}$.
Then $C_1 = \emptyset = C_2$, and also $C_1' = \emptyset$. The identification of the nodes in $V_H \cup V_{H_1} \cup V_{H_2}$ which is caused by $att_H(x_1)$ and $glue_1$ resp. $att_{H_1}(x_2)$ and $glue_2$ can be done in any order, yielding always the same hypergraph $H' = H''$. Similarly, C' and C''' resp. $glue'$ and $glue''$ are adjustments of C resp. $glue$ to the resulting quotient node set, and therefore the same. □

Now hyperedge substitution with BAR languages can be investigated.

5.26 Theorem (closure under hyperedge substitution)
AR languages and BAR languages are closed under hyperedge substitution with BAR languages.

Proof. Let $AG = (N, T, P, Z)$ be an AR grammar and, for some $\Gamma \subseteq T$, $subst: \Gamma \rightarrow [\mathcal{BMG}_{\emptyset,\emptyset,T}]$ a hyperedge substitution such that $subst(\gamma)$ is generated by a BAR grammar $AG_\gamma = (N_{V_\gamma} \uplus N_{E_\gamma}, T, P_\gamma, (Z_\gamma, \emptyset, glue_\gamma))$. Moreover, let $\hat{\Gamma} = \{\hat{\gamma} \mid \gamma \in \Gamma\}$ be another set of labels. Without loss of generality, $\hat{\Gamma}$ and the sets of nonterminal labels of the grammars may be assumed to be pairwise disjoint.

Now construct an AR grammar $AG' = (N', T, P', Z')$ from AG and the BAR grammars which, whenever a derivation in AG generates a sentential form containing a γ-labelled hyperedge with $\gamma \in \Gamma$ and whose incident nodes are terminal, generates the same sentential form but labelling that hyperedge with the new nonterminal label $\hat{\gamma}$. Moreover, add productions which can rewrite such a $\hat{\gamma}$-labelled hyperedge into the axiom $(Z_\gamma, \emptyset, glue_\gamma)$ of AG_γ. Then AG' generates $subst(L(AG))$.

Formally, let $N' = N \uplus \hat{\Gamma} \uplus \uplus_{\gamma \in \Gamma}(N_{V\gamma} \uplus N_{E\gamma})$. Moreover, let $\bar{P}$ be a set containing the productions $X ::= (H', C', glue)$ obtained from the productions $(X ::= (H, C, glue)) \in P$ by changing, in H, each label $\gamma \in \Gamma$ of a hyperedge whose incident nodes are terminal into the nonterminal label $\hat{\gamma}$, and analogously changing $lab(cr)$ to $\widehat{lab(cr)}$, for every connection instruction $(ex/cr) \in C$ which potentially generates embedding hyperedges labelled in Γ and with terminal incident nodes only. Furthermore, let $\hat{P} = \{p_\gamma \mid \gamma \in \Gamma$ and $p_\gamma = (\hat{\gamma} ::= (Z_\gamma, \emptyset, glue_\gamma))\}$. Then $P' = \bar{P} \uplus \hat{P} \uplus \uplus_{\gamma \in \Gamma} P_\gamma$ is the new production set. Finally, Z' is obtained from Z by changing the label of every γ-labelled hyperedge with exclusively terminal incident nodes into $\hat{\gamma}$ if $\gamma \in \Gamma$.

The inclusion $subst(L(AG)) \subseteq L(AG')$ can be seen as follows. Let $[H] \in L(AG)$, $E_H|_\Gamma = \{e \in E_H \mid lab_H(e) \in \Gamma\}$ and, for each hyperedge $e \in E_H|_\Gamma$, $[(H_e, \emptyset, glue_e)] \in L(AG_{lab_H(e)})$ such that H and the H_e are pairwise disjoint. This means that there are derivations $Z \Rightarrow_P^* H$ in AG and $(Z_{lab_H(e)}, \emptyset, glue_{lab_H(e)}) \Rightarrow_{P_{lab_H(e)}}^* (H_e, \emptyset, glue_e)$ in $AG_{lab_H(e)}$ (for each $e \in E_H|_\Gamma$) with all hypergraphs occurring in these derivations pairwise disjoint. Moreover, there is a derivation $Z' \Rightarrow_{\bar{P}}^* H'$ in AG' corresponding one-to-one to the derivation of H in AG, and H' differs from H only in that every hyperedge in $E_H|_\Gamma$ is labelled $\widehat{lab_H(e)}$ in H'. Continue this derivation by applying, to each hyperedge $e \in E_H|_\Gamma$, $p_{lab_H(e)}$, and then applying the productions as in the derivation $(Z_{lab_H(e)}, \emptyset, glue_{lab_H(e)}) \Rightarrow_{P_{lab_H(e)}}^* (H_e, \emptyset, glue_e)$. By the associativity of AG_γ, this has the same effect as directly substituting $(H_e, \emptyset, glue_e)$ for e. Consequently, $subst([H]) \subseteq \{[H''] \in [\mathcal{M}_T] \mid H' \Rightarrow_{\bar{P}}^* H''\}$.

For the other inclusion, consider any derivation $Z' \Rightarrow_{P'}^* H''$ in AG' with $[H''] \in L(AG')$. As every hyperedge e which is rewritten with a production in $\hat{P}$ has a border of terminal incident nodes, the replacements of e and its descendants can be shifted to the end of the derivation without changing the resulting hypergraph. Performing these shifts exactly once for each of these hyperedges, the original derivation is transformed into a derivation

$$Z' \Rightarrow_{\bar{P}}^* H' \Rightarrow_{\hat{P}} H_1 \Rightarrow_{P_{\gamma_1}}^* H_1' \Rightarrow_{\hat{P}} \cdots \Rightarrow_{\hat{P}} H_i \Rightarrow_{P_{\gamma_i}}^* H_i' \Rightarrow_{\hat{P}} \cdots \Rightarrow_{P_{\gamma_n}}^* H_n' = H''.$$

For the derivation $Z' \Rightarrow_{\bar{P}}^* H'$, there is a corresponding derivation $Z \Rightarrow_P^* H$ in AG where H and H' are related as above, and therefore $[H] \in L(AG)$. Then once more by the associativity of the grammars AG_γ, $[H''] \in subst([H])$.

Finally, if $AG = (N_V \uplus N_E, T, P, Z)$ is a BAR grammar, then let $AG' = (N_V' \uplus N_E', T, P', Z')$ with $N_V' = N_V \uplus \uplus_{\gamma \in \Gamma} N_{V\gamma}$, $N_E' = N_E \uplus \uplus_{\gamma \in \Gamma} N_{E\gamma} \uplus \hat{\Gamma}$, and the rest as above. Then it is immediate that AG' is a BAR grammar, and it generates $subst(L(AG))$. $\square$

Theorem 5.26 allows to construct many interesting graph and hypergraph languages from very simple languages in a structured way. Consider e.g. the set L containing the two basic hypergraphs with gluing

$$[(a\underset{u_1}{\bullet}\quad\underset{u_2}{\bullet}a, \emptyset, \{(aa/u_1u_2)\})] \qquad \text{and} \qquad [(a\underset{u_1}{\bullet}\overset{a}{\longrightarrow}\underset{u_2}{\bullet}a, \emptyset, \{(aa/u_1u_2)\})],$$

and the substitution $subst\colon \{a\} \to L$. Extending $subst$ to the set COMPLETE of complete a-labelled graphs, which is generated by the linear BAR grammar AG_1 from Example 5.7.1, yields the set $[\mathcal{G}_{\{a\}}]$ of all (a-labelled) graphs. It is, moreover, possible to generate with a linear edNCE grammar the set HAMILTONIAN$_C$ of complete graphs containing a directed hamiltonian cycle of b-labelled edges in an otherwise a-labelled graph. Similarly, there is a confluent edNCE grammar generating the set SPANNING$_C$ of complete graphs where b-labelled edges distinguish a(n undirected) spanning tree. Extending $subst$ to these sets yields the set HAMILTONIAN of all hamiltonian graphs resp. the set CONNECTED of all connected graphs. If desired, $subst$ can also be modified to substitute a-labelled edges for the b-labelled edges, so that the hamiltonian cycles resp. the spanning trees are no longer distinguished.

As a last example, consider the set 2-TOURNAMENTS$_0$ of all ordered tournaments, which can be generated by a linear hNCE grammar, as proposed in Example 2.9.4 (in this case, the grammar is an edNCE grammar). Substituting for the edges of these graphs one of the two basic graphs with gluing

$$[(\underset{u_1}{\bullet}\longrightarrow\underset{u_2}{\bullet}, \emptyset, \{(**/u_1u_2)\})] \qquad \text{and} \qquad [(\underset{u_1}{\bullet}\longleftarrow\underset{u_2}{\bullet}, \emptyset, \{(**/u_1u_2)\})]$$

yields the set TOURNAMENTS of all tournaments. Furthermore, using the substitution $subst$ from above (with $*$ instead of a) produces the set ACYCLIC of all acyclic graphs, which can be seen as follows: Clearly, an ordered tournament is acyclic, and removing edges does not introduce cycles. Conversely, if G is a graph containing two nodes v_1, v_2, but neither the edge $e_1 = (v_1, v_2)$ nor the edge $e_2 = (v_2, e_1)$, and if both G plus e_1 and G plus e_2 contain a cycle, then there is already a cycle in G. Consequently, given an acyclic graph, one can repeatedly add an edge until it is an acyclic graph where every two distinct nodes are adjacent. The resulting graph is a tournament because in an unlabelled graph, more than one edge between any two nodes implies a cycle. Moreover, it is ordered: considering the order on the nodes induced by the directed edges, the acyclicity implies that there is a minimal node, which must be unique because there is an edge between any two distinct nodes. As the graph without the minimal node and its incident edges is either empty or once more an ordered tournament, repeating the argument yields the required sequence of (minimal) nodes.

In the examples listed above, the edNCE grammars generating the underlying graph languages are confluent, and so are their induced BAR grammars. Thus, the languages $[\mathcal{G}]$, HAMILTONIAN, CONNECTED, and TOURNAMENTS can be generated by confluent (and associative) BAR grammars. However, these grammars are *not* context-free: the edges which are later substituted are generated as nonterminal embedding edges from terminal edges, and the number of nonterminal edges thus introduced in one derivation step is not constant. Consider for example the beginning of the derivation shown in Figure 5.6: the first application of p_1 (to the axiom) results in one nonterminal node and no nonterminal hyperedges, while the second produces, in addition to the nonterminal node, two nonterminal hyperedges. Hence, BAR grammars are not nonterminal-preserving.

5.4   Concluding Remarks

Atom replacement [HK98] combines and unifies the concepts of node replacement and hyperedge replacement. It allows to generate a large number of hyperedges in a single step, as well as to delete or transform single hyperedges in a direct way. It is more powerful than hyperedge replacement, while the exact relationship to node replacement is still open. In particular, the usefulness of combining embedding and gluing in one rule application to a node remains to be investigated beyond its being a special case of the approach in [KR90]; a convincing example would be helpful. Moreover, the relation of the approach proposed in [Bar99, Section 2.2.2], which also combines node rewriting and hyperedge rewriting, should be studied.

One of the motivations to introduce atom replacement was to define a 'most general' context-free hypergraph-rewriting approach. In general, however, atom-replacement grammars do not have any of the properties required for context-freeness in the sense of [Cou87]. Imposing associativity (and confluence) still yields an attractive hypergraph-generating device, as witnessed by BAR grammars. In contrast, as the requirement to preserve nonterminals annihilates the possibility to generate an arbitrarily large number of nonterminal embedding hyperedges in one derivation step, it seems that for AR grammars with this property, a uniquely node-rewriting normal form can be constructed analogously to the simulation of hyperedge rewriting by node rewriting, see Theorem 3.19. Under this conjecture, context-free atom-replacement has a higher generative power than confluent node rewriting only if this is caused by the possibility to glue nodes in a derivation step where a node is replaced, and it may well be that this is not the case for confluent and associative node-rewriting AR grammars.

The concept of BAR grammars, i.e. the combination of node rewriting steps with hyperedge rewriting steps in one grammar such that the grammar remains associative, is interesting in its own right. As seen at the end of the previous section,

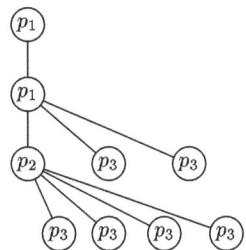

Figure 5.7: A derivation tree for the derivation of Figure 5.6

even the restriction to confluent BAR grammars, C-BAR grammars for short, does not yield a context-free rewriting device, because C-BAR grammars are not necessarily nonterminal preserving. While in [Cou87], devices which lack this property are not called grammars, the example of C-BAR grammars suggests a more general viewpoint. In fact, due to the confluence and associativity of these grammars, it should be possible to define derivation trees in a meaningful way (at least for BAR grammars where the hypergraphs do not contain adjacent nonterminal nodes). An example for such a tree, gleaned from the derivation of Figure 5.6 and built similarly to the (p-labelled) derivation trees of [ER97], is depicted in Figure 5.7: the vertices are labelled with the productions applied in the derivation, and a vertex has a child if the application of its label yields the nonterminal atom which is rewritten with the label of the child. In this example, the number of children which a (p_1-labelled) vertex has depends on the depth of the vertex in the tree; there are other C-BAR grammars where further factors play a role. The types of derivation trees occurring in C-BAR grammars may lead to a classification of their generated languages, and it should be interesting to see which derivation-tree based results known for context-free grammars carry over to confluent and associative grammars.

It may be possible to find a second classification of BAR languages which is linked to their closure under hyperedge substitution. Owing to the associativity of BAR grammars, the effect of rewriting a hyperedge and subsequently all its nonterminal descendants can also be achieved by hyperedge substitution. Hence, a BAR derivation can be seen as a number of node-rewriting steps, followed by a hyperedge substitution, followed again by node-rewriting steps, and so forth. In other words, a BAR language can be generated by an iteration of hyperedge substitutions with node-rewriting languages, and there may be a hierarchy of BAR languages based on the number of iterations it takes minimally to generate a certain language.

A further perspective on atom replacement is given by viewing it in the context of the graph- and rule-centred programming language GRACE [AEH+99, KK99, Kus00b]. As a unification of a node-rewriting and a hyperedge-rewriting approach to (hyper)graph rewriting, it is an example of a hybrid approach. For example, one might implement the extension of a hyperedge substitution to a node-rewriting language by writing one transformation unit to generate this language, and one unit, for each symbol, to generate the language which is assigned to this symbol by the substitution. Choosing in the latter units all hypergraphs as initial, a further transformation unit which imports the others then produces the substituted language. Moreover, the alternation between node rewriting and hyperedge rewriting/substitution can be steered by control conditions such as the ones presented in [Kus00a].

6

Application
of Node and Hyperedge Rewriting
to Net Refinement

Petri nets are well known as models for nondeterministic concurrent systems. For the structured top-down design of large Petri net models, the notion of net refinement has been introduced, see e.g. [GSW80, Rei87]. Two main aspects of refinement are considered in the literature, cf. the survey in [BGV91]: A refinement can preserve behaviour equivalence, i.e. it transforms semantically equivalent coarse nets into semantically equivalent refined nets (see e.g. [GG90, Vog91, Vog97]), or it can be behaviour-preserving, i.e. the behaviour of the refined net can be inferred from the behaviour of the coarse net (see e.g. [SM83, Vog87, vdA97]). In either case, iterating refinement steps on a net allows to view the model at different levels of abstraction, thus yielding a notion of hierarchical Petri nets such as in [Feh93].

The net structure underlying a Petri net is usually defined as a bipartite graph, but can also be seen as a hypergraph. Thus, refinement is basically a (hyper)graph transformation, and it may be context-free if only a small part of a net can be changed in one refinement step. The derivation trees obtained in that case correspond in a natural way to hierarchical Petri nets.

In this chapter, two particular net refinement techniques taken from the literature are shown to be examples for context-free hypergraph rewriting techniques. Both of them refine transitions, and both are behaviour-preserving: they preserve liveness and safeness. While the technique van Glabbeek and Goltz present in [GG90] uses a method to connect the refinement net to the context net of the refined transition which is best modelled by node rewriting, van der Aalst's technique for modelling workflow processes [vdA97] validates the concept of grammars in this context.

Section 6.1 contains a brief introduction to Petri nets. The general idea of net refinement and of its implementation as (hyper)graph rewriting are given in Section 6.2. The refinement techniques mentioned above are treated in Sections 6.3 and 6.4. Finally, an evaluation of the situation is given in Section 6.5.

6.1   Net Structures and Petri Nets

The brief introduction to Petri nets given in this section follows the concise presentation in [Mur89]. A general introduction can be found in e.g. [Rei85].

A *net structure* is a triple $N = (P_N, T_N, F_N)$ where P_N and T_N are disjoint finite sets of *places* resp. *transitions* and $F_N \subseteq (P_N \times T_N) \cup (T_N \times P_N)$ is a set of *arcs* (the *flow relation*). Thus, a net structure is basically an unlabelled bipartite graph. In a picture of a net structure, a place is represented by a circle $\bigcirc$, a transition by a square $\square$, and an arc (x, y) by an arrow from x to y. Figure 6.1 shows a sample net structure.

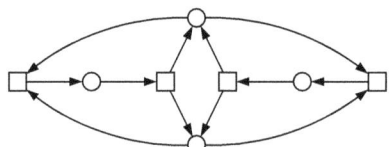

Figure 6.1: A net structure

Let $N = (P_N, T_N, F_N)$ be a net structure. For an item $x \in P_N \cup T_N$, the *preset* of x is $^\bullet x = \{y \in P_N \cup T_N \mid (y, x) \in F_N\}$, and $x^\bullet = \{y \in P_N \cup T_N \mid (x, y) \in F_N\}$ is its *postset*. If x is a transition, $^\bullet x$ contains the *preplaces* and $x^\bullet$ the *postplaces* of x (and analogously for a place x and its pre- and posttransitions). Moreover, the set $^\circ N = \{x \in P \mid {}^\bullet x = \emptyset\}$ contains the *initial* places of N, and $N^\circ = \{x \in P \mid x^\bullet = \emptyset\}$ its *terminal* places.

Given a net structure $N = (P, T, F)$ with $^\circ N \neq \emptyset$ and $N^\circ \neq \emptyset$, we will sometimes consider the net structure N^* obtained from N by adding a new transition t^*, arcs from t^* to every place in $^\circ N$, and arcs from every place in N° to t^*. Formally, $N^* = (P^*, T^*, F^*)$ with $P^* = P$, $T^* = T \uplus \{t^*\}$, and $F^* = F \cup \{(t^*, p) \mid p \in {}^\circ N\} \cup \{(p, t^*) \mid p \in N^\circ\}$. The construction is sketched in Figure 6.2.

Let $N = (P_N, T_N, F_N)$ be a net structure. A *marking* of N is a mapping $M: P_N \to \mathbb{N}$ assigning $M(p)$ *tokens* to each place $p \in P_N$. It is depicted by drawing $M(p)$ dots into the circle for $p \in P_N$.

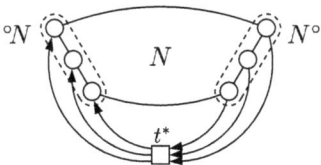

Figure 6.2: Constructing N^* from N

A(n ordinary) *Petri net* is a pair $PN = (N, M_0)$ where N is a net structure and M_0 is a marking of N called the *initial* marking. Figure 6.3 shows a Petri net consisting of the net structure of Figure 6.1 with a(n initial) marking.

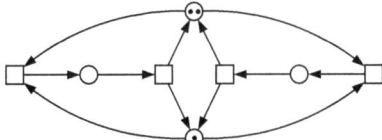

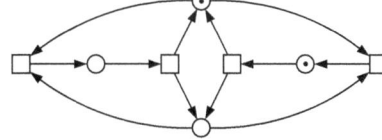

Figure 6.3: A Petri net

Figure 6.4: A follower marking of the net in Figure 6.3

Let M be a marking of a net structure $N = (P_N, T_N, F_N)$. A transition $t \in T_N$ is said to be *enabled* (denoted $M[t>$) if $M(p) > 0$ for all places $p \in {}^\bullet t$. If a transition $t \in T_N$ is enabled, it may *fire*, meaning that a token is removed from each of its preplaces and a token is added to each of its postplaces. The firing of a transition t transforms a marking M into the *follower* marking M' (denoted $M[t> M')$, where for all $p \in P_N$:

$$M'(p) = \begin{cases} M(p) - 1 & \text{if } p \in {}^\bullet t \setminus t^\bullet, \\ M(p) + 1 & \text{if } p \in t^\bullet \setminus {}^\bullet t, \\ M(p) & \text{otherwise.} \end{cases}$$

Figure 6.4 shows the Petri net of Figure 6.3 with the follower marking reached after the rightmost transition has fired.

A marking M' is *reachable* from a marking M if there is a *firing sequence* $M = M_1 [t_1> M_2 [t_2> \ldots M_{n-1} [t_{n-1}> M_n = M'$ (also denoted $M[*> M')$. The set of all markings reachable from M_0 in a Petri net (N, M_0) is denoted by $R_N(M_0)$.

A Petri net (N, M_0) is *k-bounded* or simply *bounded* if there is a natural number $k \in \mathbb{N}$ such that $M(p) \leq k$ for all $p \in P_N$ and $M \in R_N(M_0)$; a 1-bounded Petri net is also called *safe*. It is *live* if for every marking $M \in R_N(M_0)$ and every transition $t \in T_N$ there is a marking $M' \in R_N(M)$ enabling t.

The Petri net of Figure 6.3 (and Figure 6.4) is both live and bounded, but not safe.

6.2 Rule-based Net Refinement

This section provides the basis to implement net refinement as (hyper)graph rewriting. Net refinement is defined and a first system of refinement rules is given. Then three possibilities to represent Petri nets as graphs or hypergraphs are introduced. On the basis of these representations, node rewriting, hyperedge rewriting, and handle rewriting can be interpreted as net refinement. Finally, an appropriate implementation of the refinement rules previously given is discussed.

The general idea for net refinement is to expand an item—a place resp. a transition—into a larger net which is linked by arcs to the items in the pre- and postset of the refined item. Consider for example the Petri net (N, M_0) in Figure 6.5 which models a situation of mutual exclusion, with the place p acting as semaphore. The refined net (N', M_0') represents explicitly critical sections and initialisation of common resources. As indicated by the dotted lines, the operation performed here can be seen as a parallel refinement of p and its two pre- resp. posttransitions. Note that each of the three items is refined to a net whose border with respect to the whole net consists of items of the same kind as the coarse item.

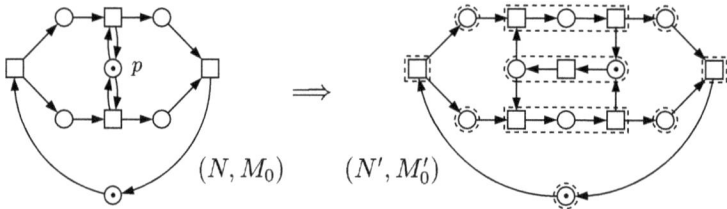

Figure 6.5: Refining a Petri net

One way to formalise net refinement is based on net morphisms [GSW80]. The equivalent formalisation given in [Rei87] is based on the abstraction of a subnet into one item, where the border of the subnet with respect to the whole net consists entirely of one kind of items, and the item into which the subnet is contracted is of the same kind. We choose the former method because net morphisms will be needed later to study net refinement in the context of pullback rewriting.

6.1 Definition (net morphism)
Let $N = (P, T, F)$ and $N' = (P', T', F')$ be two net structures. A *net morphism* $f: N \to N'$ is a mapping $f: P \cup T \to P' \cup T'$ satisfying, for all $(x, y) \in F$ with $f(x) \neq f(y)$, $(f(x), f(y)) \in F'$ and $x \in P \Leftrightarrow f(x) \in P'$. If f is bijective and both f and f^{-1} are net morphisms, then f is a *net isomorphism* and N, N' are called *isomorphic*.

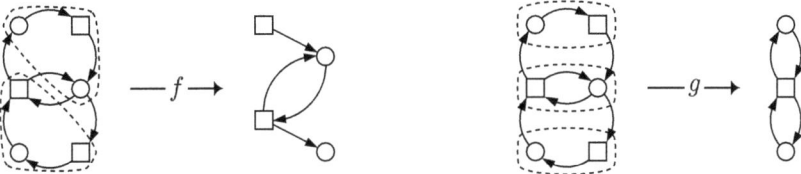

Figure 6.6: A net morphism   Figure 6.7: No net morphism

The mapping f shown in Figure 6.6 which maps the lower group of items to the lower transition and the upper group to the upper place is a net morphism. In contrast, the mapping g of Figure 6.7 is no net morphism because it does not preserve the types of the items connected by the downward arcs. Note that transitions may be mapped to places and vice versa, as long as the border of a subnet mapped to one item is of the same kind as that item.

A special type of net morphisms captures the idea of net refinement as discussed above.

6.2 Definition (refinement)
Let $N = (P, T, F)$ and $N' = (P', T', F')$ be two net structures. A *quotient* is a net morphism $f \colon N \to N'$ which is surjective with respect to items as well as arcs, i.e. for all $x' \in P' \cup T'$ there is $x \in P \cup T$ with $f(x) = x'$, and $(f^{-1}(x') \times f^{-1}(y')) \cap F \neq \emptyset$ for all $(x', y') \in F'$. Given such a quotient, N is called a *refinement* of N', and N' is an *abstraction* of N.

While in Figure 6.5, the net structure N' is a refinement of the net structure N, the net morphism f of Figure 6.6 does not specify a refinement because, in the target net, the upper transition, the lower place, and their arcs do not have inverse images under f. Note, moreover, that the definition of a refinement is based solely on net structures; we will speak of a (Petri) net refinement if the operation is a refinement with respect to the underlying net structures.

In the literature, various types of refinement rules are studied. For our purposes, we are interested in refinement rules where a place or, more commonly, a transition can be refined without checking for the presence or absence of some context, and where the refinement operation affects only a 'small' context of the refined item (e.g. only its pre- and postset).

6.3 Example (simple refinement rules)
Consider the six net transformation rules in Figure 6.8, taken from [Mur89, Section V.C] and previously presented for marked graphs in [MK80]; they can be shown to preserve liveness, boundedness, and safeness. Read from left to right, they are abstraction or *reduction* rules: (a) fusion of series places, (b) fusion of series transitions, (c) fusion of parallel places, (d) fusion of parallel transitions, (e) elimination

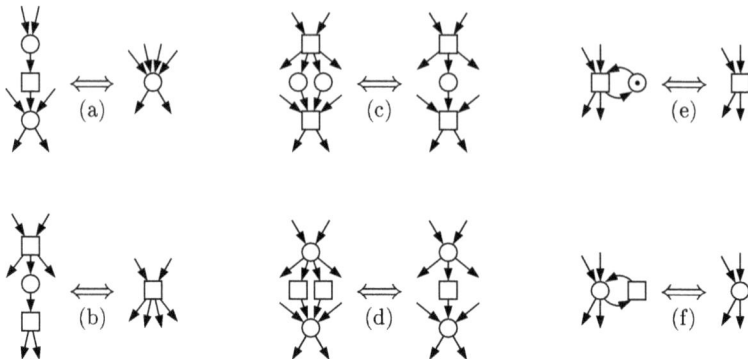

Figure 6.8: Net transformation rules preserving liveness and safeness

of self-loop places, (f) elimination of self-loop transitions. Read from right to left, they are refinement rules in the sense above, introducing resp. adding that which is fused resp. eliminated the other way round. Intuitively, it is clear that the arcs where either the origin or the target is not represented specify which type of context may (but need not) occur to allow the application of a rule. For example, the transitions in rule (c) may have any number of in- or outgoing arcs, and these arcs are invariant under an application of this rule, but the places may have arcs only as specified explicitly. With this interpretation, the reduction rules are deterministic, but not all of the refinement rules are: Consider e.g. refinement rule (b), which is the rule applied to the two refined transitions of the left net in Figure 6.5. The rule specifies that arcs going in to the unrefined transition are transferred to the first (upper) transition of the refining net, whereas arcs going out of the unrefined transition may go out from either of the refining transitions. (Analogously, refinement rule (a) is not deterministic.) Thus, the refinement in Figure 6.5 illustrates but one possible choice. Finally, note that the refinement of place p in Figure 6.5 is not achieved by applying rule (a) because this rule is applicable to unmarked places only and p contains a token.   ∎

Let us now turn to the relationship between net refinement and hypergraph rewriting. As a first step, a translation of nets into graphs or hypergraphs is needed.

There are three basic possibilities to translate net structures: into bipartite graphs, into hypergraphs with places represented by nodes, and into hypergraphs with places represented by hyperedges. Usually, net structures are introduced as bipartite graphs, which suggests the first translation. The following sections, however, make clear that sometimes the other translations are more adequate.

In addition, the marking of a place can be encoded into the label of the representing atom. In order to keep this as simple as possible, we will restrict our attention to a special case:

General assumption. For the rest of this chapter, only safe Petri nets and safeness-preserving net refinements are considered.

Formally, the three translations of Petri nets read as follows.

6.4 Definition (translations of Petri nets)

Let $\Sigma = \Delta \uplus \Delta_\circ \uplus \Delta_\bullet$ be a set of symbols, where symbols in $\Delta_\circ = \{\alpha_\circ \mid \alpha \in \Delta\}$ are used to label unmarked 'places', symbols in $\Delta_\bullet = \{\alpha_\bullet \mid \alpha \in \Delta\}$ to label marked 'places', and symbols in Δ to label 'transitions.' Moreover, associate with each $\alpha \in \Sigma$ a natural number $n_\alpha \in \mathbb{N}$ and a *direction function* $d_\alpha \colon [n_\alpha] \to \{in, out\}$. Now, let (N, M_0) be a Petri net with $N = (P, T, F)$.

Translation 1: Nets are bipartite graphs, where places as well as transitions are nodes, and arcs are edges. The Petri net (N, M_0) is encoded as a graph (V, E, lab) where $V = P \cup T$, $E \subseteq \Delta \times F$ with, for each $(x, y) \in F$, exactly one $\alpha \in \Delta$ such that $(\alpha, x, y) \in E$, $lab(p) \in \Delta_\circ$ for each $p \in P$ with $M_0(p) = 0$, $lab(p) \in \Delta_\bullet$ for each $p \in P$ with $M_0(p) = 1$, and $lab(t) \in \Delta$ for each $t \in T$.

Translation 2: Nets are hypergraphs, where places are nodes and transitions are hyperedges. The Petri net (N, M_0) is encoded as a hypergraph (V, E, lab, att) where $V = P$, $E = T$, $lab(p) \in \Delta_\circ$ for each $p \in V$ with $M_0(p) = 0$, $lab(p) \in \Delta_\bullet$ for each $p \in V$ with $M_0(p) = 1$, and for each $t \in E$, $lab(t) \in \Delta$ with $n_{lab(t)} = \#^\bullet t + \#t^\bullet = rank(t)$, and $att(t)$ is a repetition-free sequence over ${}^\bullet t \cup t^\bullet$ such that $d_{lab(t)}({}^\bullet t) = \{in\}$ and $d_{lab(t)}(t^\bullet) = \{out\}$.

Translation 3: Nets are hypergraphs, where places are hyperedges and transitions are nodes. The Petri net (N, M_0) is encoded as a hypergraph (V, E, lab, att) where $V = T$, $E = P$, $lab(t) \in \Delta$ for each $t \in V$, and for each $p \in E$, $lab(p) \in \Delta_\circ$ if $M_0(p) = 0$, $lab(p) \in \Delta_\bullet$ if $M_0(p) = 1$, $n_{lab(p)} = \#^\bullet p + \#p^\bullet = rank(p)$, and $att(p)$ is a repetition-free sequence over ${}^\bullet p \cup p^\bullet$ such that $d_{lab(p)}({}^\bullet p) = \{in\}$ and $d_{lab(p)}(p^\bullet) = \{out\}$.

It should be obvious that each translation is an encoding, since the net can be reconstructed from the graph or hypergraph which represents it. In fact, the graph or hypergraph contains, encoded in the labels resp. tentacle numbers, more information than the net itself; this proves to be useful to model certain refinement operations. Finally, note that Translations 2 and 3 may produce multiple hypergraphs if the net to be translated contains two distinct transitions resp. places with identical pre- and postsets.

6.5 Example (translations of Petri nets)

Consider the net structure of Figure 6.1 which, taking up Dijkstra's Dining Philosophers [Dij71], represents two dining philosophers (the left and right places) and two
forks (the upper and lower places). Moreover, a philosopher can alternate between
taking up two forks and eating (the outermost transitions) and putting the forks
back to start thinking (the innermost transitions).

With Translation 1, this net structure can be represented by the graph of Figure 1.1, where the label P refers to a philosopher, f to a fork, e to eating, t to
thinking, l to left, and r to right. Moreover, this translation allows to interpret
the edNCE rewriting step of Example 2.2 as a place refinement where a used fork
is replaced with two fresh ones between which a third philosopher is seated. In
this rewriting step, the labels as well as the directions of the edges incident to the
rewritten node are used to distinguish the respective neighbours.

With Translation 2, the net structure can be represented by the hypergraph
of Figure 1.2. Here, $d_e(1) = d_e(2) = in = d_t(1)$ and $d_e(3) = out = d_t(2) =$
$d_t(3)$, i.e. the mappings d_e and d_t represent the 'direction' of tentacles. Then, the
hNCE rewriting step of Example 2.5 may be seen as a place refinement where
one philosopher fetches two fresh forks and invites two of her colleagues to sit at
her right resp. left. Based on the same translation, Example 3.23 expresses the
refinement of the thinking process, including all its resources (one philosopher and
two forks), by two philosophers, three forks, and the appropriate thinking and
eating processes.

With Translation 3, the net structure can be represented by the uppermost
hypergraph in Figure 3.3. Here, $d_f(2) = d_f(3) = in = d_P(2)$ and $d_f(1) = d_f(4) =$
$out = d_P(1)$. Based on this translation, Example 3.13 is an alternative implementation of a place refinement where a fork is removed and one philosopher and two
forks placed in its stead.   ∎

It should be noted that Habel defines hypergraphs in [Hab92a] such that hyperedges have, instead of one sequence of attachment nodes, two sequences of source
resp. target nodes. Clearly, this makes associating a direction function d_α with
every label α superfluous. On the other hand, defining context-free hypergraph
rewriting approaches becomes more involved, without a gain in expressive power
(see e.g. Corollary 4.7 in [Hab92a, Chapter I]).

Note, moreover, that Kreowski proposes in [Kre81] an alternative translation of
Petri nets to (unlabelled) graphs where a token is represented by a node which is
adjacent to the 'place' carrying that token. For example, the graph in Figure 6.9
represents the Petri net of Figure 6.3. While, as shown in that reference, such
a translation is serviceable to encode the firing of a transition as the application
of a rule (in the double-pushout approach to graph rewriting [Ehr79]), there are
refinements of marked places which cannot be modelled in a context-free way.

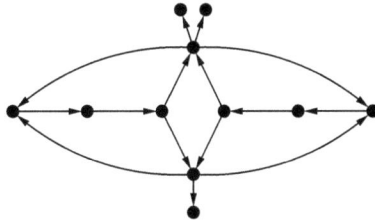

Figure 6.9: An alternative encoding of Petri nets

Let us now discuss how the refinement rules of Example 6.3 can be implemented as (hyper)graph rewriting productions.

6.6 Example (implementation of refinement rules)
It is suitable to implement the system of refinement rules presented in Example 6.3 as a system of edNCE rewriting rules, based on Transformation 1. This is because, as there are both place and transition refinement rules in the system, any implementation based on Translation 2 or 3 would have to use a combination of node and hyperedge replacement, and thus be more complex than necessary.

Let the alphabet Δ contain labels $*$, 1, and 2. We show only the implementation of rules (a), (c), and (e); the other rules can be treated analogously. The left-hand side of a production will determine whether it can be applied to an 'unmarked place' (label $*_o$) or to a 'transition' (label $*$).

The following edNCE production implements refinement rule (a) of Figure 6.8, where p_1 is the top node and p_2 the bottom node of the right-hand side graph, ingoing 1-labelled edges are transferred to p_1, and ingoing 2-labelled resp. outgoing ($*$-labelled) edges are transferred to p_2:

$$*_o ::= \left(\begin{array}{c} \bullet\, {}^{*_o} \\ \bullet\, {}^{*} \\ \bullet\, {}^{*_o} \end{array} \,,\, \left\{ \begin{array}{l} (1, *, in/*, p_1, in), \\ (2, *, in/*, p_2, in), \\ (*, *, out/*, p_2, out) \end{array} \right\} \right)$$

Refinement rule (c) is implemented by the following edNCE production, where p_1 is the left and p_2 the right node of the right-hand side graph:

$$*_o ::= \left(\begin{array}{cc} \bullet & \bullet \\ {}^{*_o} & {}^{*_o} \end{array} \,,\, \left\{ \begin{array}{l} (*, *, in/*, p_1, in), (*, *, out/*, p_1, out), \\ (*, *, in/*, p_2, in), (*, *, out/*, p_2, out) \end{array} \right\} \right)$$

Refinement rule (e) is implemented by the following edNCE production, where t is the left node of the right-hand side graph:

$$* ::= \left(\begin{array}{cc} \bullet\!\!\rightarrow\!\!\bullet \\ {}^{*} \quad {}^{*_\bullet} \end{array} \,,\, \left\{ \begin{array}{l} (*, *_o, in/*, t, in), (*, *_o, out/*, t, out), \\ (*, *_\bullet, in/*, t, in), (*, *_\bullet, out/*, t, out) \end{array} \right\} \right)$$

Note that the implementation of rule (a), and analogously that of rule (b), makes use of edge labels 1 and 2 to distinguish between edges having the same direction which are to be transferred to different nodes of the right-hand side graph. Clearly, if in a net there is an arc from a transition to a place, the place is refined with rule (a), and the transition with rule (b), then this arc may, in the refined net, link either of two transitions with either of two places. This potential for non-confluence is reflected in the implementation of the rules as edNCE productions.

■

6.3   Refinement Nets with Initial Concurrency

From an intuitive viewpoint, transition refinement seems more natural than place refinement: A transition represents a specific task, and its refinement corresponds to specifying subtasks of which that task is composed. Particular interest has been accorded to transition refinement operations which preserve liveness and safeness (resp. boundedness), see e.g. [Val79, SM83, Vog87, GG90]. Among these, the operation by van Glabbeek and Goltz [GG90] is the only one to allow the refinement of a transition with a net which may have initial concurrency (and also terminal concurrency). This is made possible by suitably multiplying the preplaces (and also postplaces) of the refined transition. Figure 6.10 (cf. [GG90, Example 4.7]) illustrates such a refinement, where every preplace of transition t in the left net is multiplied by two (and every postplace by one), yielding the net on the right.

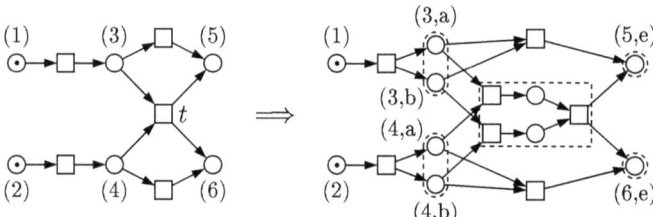

Figure 6.10: Transition refinement [GG90]

In the rest of the section, we first present the refinement operation from [GG90], and then its implementation as hypergraph rewriting.

General assumption.   In this section, only loop-free Petri nets are considered where every transition has at least one preplace and one postplace, i.e. $^\bullet t \cap t^\bullet = \emptyset$, $^\bullet t \neq \emptyset$, and $t^\bullet \neq \emptyset$ for every transition t. Moreover, Petri nets are assumed to be safe.

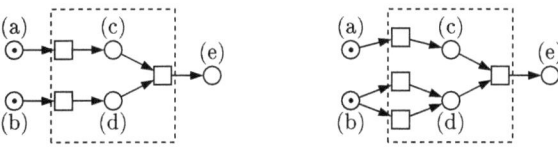

Figure 6.11: Two refinement nets

The nets which may be used to refine a transition are required to satisfy the following conditions.

6.7 Definition (refinement net)

A Petri net (N, M_0) with $N = (P, T, F)$ is a *refinement net* if

- $°N \neq \emptyset$ and $N° \neq \emptyset$,

- $M_0(p) = 1$ for all $p \in °N$ and $M_0(p) = 0$ for all $p \in N°$,

- $M_0[t>$ for any $t \in T$ only if $°t \cap °N \neq \emptyset$, and

- $M[t^*>$ for any $M \in R_{N^*}(M_0)$ only if $M[t^*> M_0$.

Figure 6.11 shows two refinement nets with initial places (a), (b) and terminal place (e). The left refinement net is used to refine the transition t in Figure 6.10: the transition is removed from the net together with its pre- and postplaces; the refinement net without its initial and terminal places is inserted; a new place (p, p') is created for every preplace p of t and every initial place p' of the refinement net, with ingoing arcs from each pretransition of p (the initial place p' does not have a pretransition) and outgoing arcs to each posttransition of p resp. p'; and analogously for the postplaces of t and the terminal places in the refinement net. This refinement operation is formally defined as follows.

6.8 Definition (net refinement)

Let (N_1, M_1) be a Petri net and t a transition in N_1. Moreover, let (N_2, M_2) be a refinement net (with N_2 disjoint from N_1). Then the refined net structure $N_3 = N_1[t/N_2]$ is defined by

- $P_3 := (P_1 \setminus (°t \cup t°)) \cup (P_2 \setminus (°N_2 \cup N_2°)) \cup Int$,
  where $Int := (°t \times °N_2) \cup (t° \times N_2°)$,

- $T_3 := (T_1 \setminus \{t\}) \cup T_2$, and

- $F_3 := ((F_1 \cup F_2) \cap (P_3 \times T_3 \cup T_3 \times P_3))$
  $\cup \{((p_1, p_2), t_1) \mid (p_1, p_2) \in Int,\ t_1 \in T_1 \setminus \{t\},\ (p_1, t_1) \in F_1\}$
  $\cup \{(t_1, (p_1, p_2)) \mid (p_1, p_2) \in Int,\ t_1 \in T_1 \setminus \{t\},\ (t_1, p_1) \in F_1\}$
  $\cup \{((p_1, p_2), t_2) \mid (p_1, p_2) \in Int,\ t_2 \in T_2,\ (p_2, t_2) \in F_2\}$
  $\cup \{(t_2, (p_1, p_2)) \mid (p_1, p_2) \in Int,\ t_2 \in T_2,\ (t_2, p_2) \in F_2\}$.

The initial marking of the refined Petri net $(N_3, M_3) = (N_1, M_1)[t/(N_2, M_2)]$ is given by

$$M_3(p) = \begin{cases} M_1(p) & \text{if } p \in P_1 \smallsetminus (\,{}^\bullet t \cup t^\bullet), \\ M_2(p) & \text{if } p \in P_2 \smallsetminus ({}^\circ N_2 \cup N_2{}^\circ), \\ M_1(p_1) & \text{if } p = (p_1, p_2) \in Int. \end{cases}$$

This refinement operation is safeness-preserving, and liveness-preserving whenever (N_2^*, M_2) is live, where (N_2, M_2) is the refinement net. Moreover, if (N_1, M_1) contains two distinct transitions t, t', and (N_2, M_2), (N_2', M_2') are refinement nets such that N_1, N_2, and N_2' are pairwise disjoint, then the two refined Petri nets $(N_1, M_1)[t/(N_2, M_2)][t'/(N_2', M_2')]$ and $(N_1, M_1)[t'/(N_2', M_2')][t/(N_2, M_2)]$ are isomorphic.

A suitable implementation of this refinement operation has to take into account the multiplication of the preplaces (resp. postplaces) of the refined transition with the initial (resp. terminal) places of the refinement net. Among the (hyper)graph rewriting techniques which we have studied in Chapters 2 and 3, only the hNCE approach and the HH approach can model such a multiplication. For both approaches, places must become hyperedges, so that Translation 3 will be used to implement nets as hypergraphs.

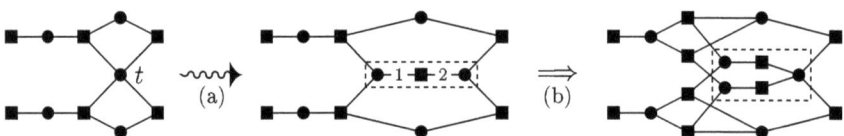

Figure 6.12: Enlarging t to a handle (1), and rewriting the handle (2)

In order to use handle replacement, it would be necessary to enlarge, prior to the actual refinement, the transition to a handle as sketched in Figure 6.12. Moreover, if the refinement net has an initial place with more than one posttransition, or a terminal place with more than one pretransition, then the refinement operation changes the 'type' of the pre- resp. postplaces of the refined transition; see for an example the refinement shown in Figure 6.13, where the right refinement net from Figure 6.11 is used. Consequently, handle rewriting cannot model this refinement operation in all possible cases.

The hNCE approach allows to specify the splitting of tentacles as well as the multiplication of hyperedges. Thus, the refinement of Figure 6.13 can be encoded as the hNCE rewriting step of Figure 6.14, where the direction functions d_i in the

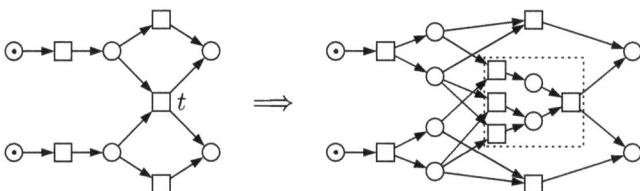

Figure 6.13: Transition refinement with the second refinement net

hyperedge labels are visualised by the directions on the tentacles, and the node label $*$ is not drawn. In this rewriting step, the node t is replaced by a hypergraph with embedding

$$
\left(
\begin{array}{c}
u_1 \bullet \xrightarrow{1} d_4 \circ \\
u_2 \bullet \xrightarrow{1} \blacksquare \xrightarrow{2}{3} \bullet u_4 \\
u_3 \bullet \xrightarrow{2} \blacksquare \\
d_5 \circ
\end{array}
, C
\right)
$$

where C contains, among others, the connection instructions $(d_{2\circ}, *\lozenge*/d_{2\circ}, 1u_13)$, $(d_{2\circ}, *\lozenge*/d_{6\circ}, 1u_2u_33)$, $(d_{3\circ}, *\lozenge/d_{3\circ}, 1u_4)$, $(d_{3\circ}, \lozenge*/d_{3\circ}, u_42)$.

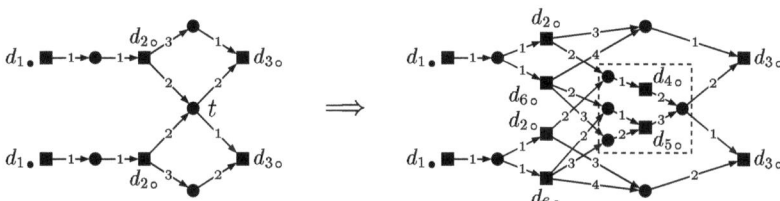

Figure 6.14: Transition refinement as node rewriting

For the formal treatment, we will consider the case of Petri nets without 'parallel' places, so that their encoding yields simple hypergraphs.

6.9 Definition

Two distinct places p_1, p_2 in a Petri net are *parallel* if $\bullet p_1 = \bullet p_2$ and $p_1^\bullet = p_2^\bullet$. A Petri net is *simple* if it does not contain parallel places.[1]

It is immediate that refining a transition in a simple Petri net with a simple refinement net yields again a simple Petri net.

---

[1]Note that in the literature, a Petri net is called simple if it does not contain *transitions* which are parallel in this sense.

Let us now turn to the encoding of refinement in simple Petri nets into hNCE rewriting in simple hypergraphs; the case of parallel places will be discussed at the end of the section.

6.10 Construction (simulation of net refinement)

Let Δ contain the symbol $*$ and every direction function $d \in \bigcup_{n \in \mathbb{N}}\{in, out\}^{[n]}$. We use Translation 3 to get a hypergraph from a simple Petri net, labelling every node of the hypergraph with $*$ and every hyperedge with the correct direction function d indexed by $\bullet$ or $\circ$ as appropriate.

Let (N, M) be a simple refinement net, and assume an arbitrary but fixed order on the transitions of N. Construct a hypergraph with embedding (H, C) from (N, M) as follows:

The hypergraph H is obtained from (N, M) by removing the initial and terminal places as well as their associated arcs from N, restricting M to the new set of places, and translating the resulting Petri net to a hypergraph as described above.

The connection relation C contains all (link-preserving) connection instructions $coin_?(d, m, p)$ such that a hyperedge with d as direction function and $d_?$ as label is transformed by replacing, in the attachment sequence, the mth node (which is the rewritten one) with the 'transitions' linked by an arc to some place $p \in {}^{\circ}N \cup N^{\circ}$. More precisely, let $d\colon [n] \to \{in, out\}$ be a direction function, $? \in \{\bullet, \circ\}$, and $m \in [n]$. Moreover, let $p \in {}^{\circ}N \cup N^{\circ}$ and $t_1, \ldots, t_k$ the transitions of N (in the given order) which are linked by an arc to p, i.e. $p \in {}^{\circ}N$ and $\{t_1, \ldots, t_k\} = p^{\bullet}$ if $d(m) = in$, and $p \in N^{\circ}$ and $\{t_1, \ldots, t_k\} = {}^{\bullet}p$ if $d(m) = out$. Then $coin_?(d, m, p) = (d_?, x_1 \ldots x_n/d'_?, y_1 \ldots y_{n+k-1})$ with $x_m = \Diamond$ and $x_i = *$ for all $i \in [n] \smallsetminus \{m\}$, $y_i = i$ and $d'(i) = d(i)$ for all $i \in \{1, \ldots, m-1\}$, $y_i = t_{i-m+1}$ and $d'(i) = d(m)$ for all $i \in \{m, \ldots, m+k-1\}$, and $y_i = i-k+1$ and $d'(i) = d(i-k+1)$ for all $i \in \{m+k, \ldots, n+k-1\}$.

Note that properly speaking, this construction does not yield a hypergraph with embedding because the connection relation is infinite. The reason for this is that repeated refinement with a refinement net such as the right one in Figure 6.11 can lead to places with arbitrarily many pre- or postplaces.

Moreover, it is easy to see from the connection relations that whenever two nodes in a net-encoding hypergraph are replaced with two refinement net-encoding hypergraphs with embedding, the result does not depend on the order in which the replacements are performed. In fact, we have here an instance of static confluence as discussed at the end of Chapter 4.

6.11 Theorem (correctness)

Let (N_1, M_1) be a simple Petri net containing a transition t and (N_2, M_2) a simple refinement net with N_2 disjoint from N_1. If H_1 is a hypergraph representing

(N_1, M_1) and (H_2, C_2) a hypergraph with embedding representing (N_2, M_2) as described in Construction 6.10, then $H_1[t/(H_2, C_2)]$ represents the simple Petri net $(N_1, M_1)[t/(N_2, M_2)]$ (up to isomorphism).

Proof. Let $H_3 = H_1[t/(H_2, C_2)]$ represent the simple Petri net (N, M), and let $(N_3, M_3) = (N_1, M_1)[t/(N_2, M_2)]$. Both (N, M) and (N_3, M_3) consist of (N_1, M_1) minus t, $^\bullet t$, and $t^\bullet$; (N_2, M_2) minus $^\circ N_2$ and $N_2{}^\circ$; and some places linked by arcs to some transitions in $(^\bullet t \cup t^\bullet \cup {}^\circ N_2)^\bullet \cup {}^\bullet(^\bullet t \cup t^\bullet \cup N_2{}^\circ)$. These places correspond as follows:

> There is, in (N_3, M_3), a ?-marked place $(p_1, p_2) \in {}^\bullet t \times {}^\circ N_2$ with pretransitions $^\bullet p_1$ and posttransitions $(p_1^\bullet \smallsetminus \{t\}) \cup p_2^\bullet$
>
> $\Longleftrightarrow$ there is, in (N_1, M_1), a ? marked place $p_1 \subset {}^\bullet t$ with pretransitions $^\bullet p_1$ and posttransitions $p_1^\bullet$, and there is a place $p_2 \in {}^\circ N_2$
>
> $\Longleftrightarrow$ there is, in H_1, a $d_?$-labelled hyperedge p_1 with ingoing tentacles from $^\bullet p_1$ and outgoing tentacles to $p_1^\bullet$ (the mth tentacle goes out to t), and there is a connection instruction $coin_?(d, m, p_2) \in C_2$
>
> $\Longleftrightarrow$ there is, in H_3, a $d_?'$-labelled hyperedge $(p_1, coin_?(d, m, p_2))$ (created from the hyperedge p_1 incident to t and the connection instruction $coin_?(d, m, p_2) \in C_2$) with ingoing tentacles from the nodes in $^\bullet p_1$ and outgoing tentacles to the nodes in $(p_1^\bullet \smallsetminus \{t\}) \cup p_2^\bullet$
>
> $\Longleftrightarrow$ there is, in (N, M), a ?-marked place $(p_1, coin_?(d, m, p_2))$ with pretransitions $^\bullet p_1$ and posttransitions $(p_1^\bullet \smallsetminus \{t\}) \cup p_2^\bullet$.

The correspondences for the places $(p_1, p_2) \in t^\bullet \times N_2{}^\circ$ are established analogously. □

For Petri nets which have parallel places (but not among initial or terminal places) and cannot be represented by simple hypergraphs, a translation analogous to Construction 6.10 but into multiple hypergraphs can be considered. It yields a substitution of multiple hypergraphs with embedding for nodes in multiple hypergraphs which works just like atom replacement with empty gluing relations; for the formal definition see Section 7.4. If, moreover, refinement nets with parallel initial or terminal places are admitted, then this concept has to be expanded once more and in such a way that connection relations are multisets instead of sets.

6.4 Refinement of Workflow Nets

While in earlier papers on net refinement, the focus lay on ever more general and powerful refinement techniques, recently some interest has been accorded to determining refinement operations suitable for some specific application. Examples for the latter comprise van der Aalst's workflow nets [vdA98] and their refinement [vdA97], and the refinement of Petri nets modelling schedules for manufacturing cells [Zub98]. In these references, refinement systems are presented which are essentially context-free grammars. In this section, that statement is made precise for the workflow net refinements by implementing them as hyperedge rewriting.

A workflow net is a Petri net specification of a workflow process. The definition of such a process is oriented at the handling of one specific case, determining the tasks that have to be executed, as well as a partial order on these tasks. For the routing of a case along the tasks, four typical situations can be distinguished:

(a) Two tasks can be accomplished sequentially;

(b) either one or the other of two tasks has to be executed;

(c) two tasks are independent and may be performed in parallel;

(d) a task may need multiple repetitions.

Petri nets are suitable to model workflow processes. More precisely, transitions, places, and tokens can be used to model, respectively, tasks, conditions on the tasks, and cases.

6.12 Example (workflow process)
Consider the process of writing a doctoral thesis. A doctoral student writing a thesis will repeatedly go through phases of intensive work. Such a phase may consist of a part where the student is working on her own, in any order studying the literature and producing text, and a part where she either talks to her supervisor or does other things. After some number of repetitions, the thesis is (hopefully) completed.

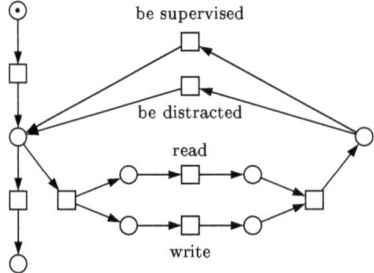

Figure 6.15: A Petri net modelling a workflow process

The Petri net in Figure 6.15 models this workflow process definition. The description of a task is written next to the transition modelling that task, and the token represents the case of one particular thesis.   ∎

The typical properties of a net modelling a workflow process are as follows.

6.13 Definition (workflow net)

A *workflow net* is a net structure N with exactly one initial and exactly one terminal place (called *input* and *output* place, respectively) such that N^*, seen as a graph, is strongly connected. The initial (resp. terminal) marking M_{in} (resp. M_{out}) assigns one token to the input (resp. output) place and no token otherwise. A workflow net is *sound* if M_{out} can be reached from every marking in $R_N(M_{in})$, M_{out} is the only marking in $R_N(M_{in})$ assigning a token to the output place of N, and each transition of N is enabled by some marking in $R_N(M_{in})$.

As shown in [vdA97], a workflow net N is sound if and only if (N^*, M_{in}) is live and bounded. Clearly, the most basic sound workflow net is the net structure N_{basic} which has, apart from one initial and one terminal place, only one transition, representing one (usually highly abstract) task.

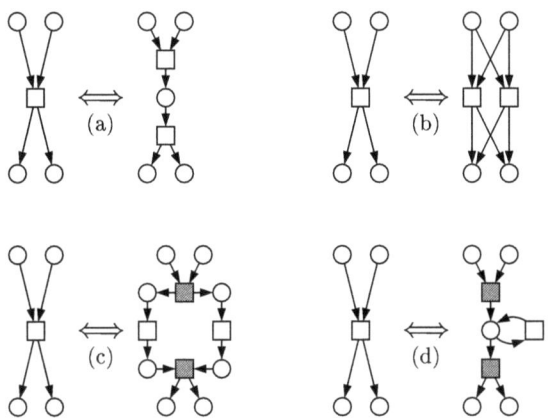

Figure 6.16: Refinement rules for workflow nets

Corresponding to the four typical routing situations mentioned above, the four transition refinement rules of Figure 6.16 are proposed in [vdA97]. Rule (a) expresses that an abstract task may consist of two tasks which have to be executed one after the other. According to rule (b), an abstract task can consist of choosing and performing one of two alternative tasks. For an abstract task consisting of two subtasks both of which have to be executed, but independently of each other, rule (c) is appropriate. Note that in addition to the (white) transitions representing the subtasks, two (grey) transitions are used which represent control activities; in this instance, a *fork* and a *join* activity. The last rule expresses the refinement of a task into an iteration of a more specific task. Again, the grey transitions represent control activities; here, the start and end of the iteration.

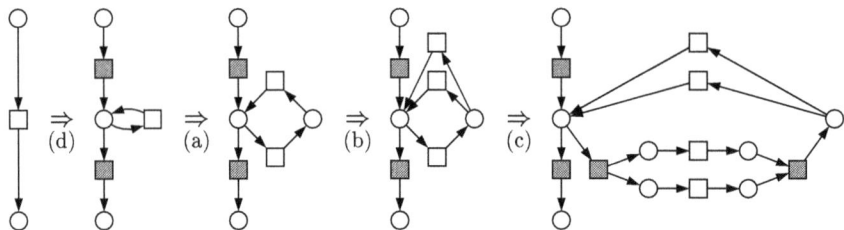

Figure 6.17: A sequence of workflow net refinements

Figure 6.17 shows how applying these rules allows to transform N_{basic} into the workflow net underlying the Petri net of Figure 6.15. The rules are interpreted similar to the rules of Example 6.3 in that they can be applied to any transition, not just transitions with exactly two pre- resp. postplaces. Together with N_{basic} as an initial workflow net, they make up a grammatical device for the generation of sound workflow nets. Assuming that transitions representing control activities may not be refined, all transitions representing tasks have exactly one pre- resp. posttransition (which may be identical, see the transition representing the iterated task in rule (d)). Then it is easy to construct a hyperedge-rewriting grammar which formalises this workflow-net refinement system.

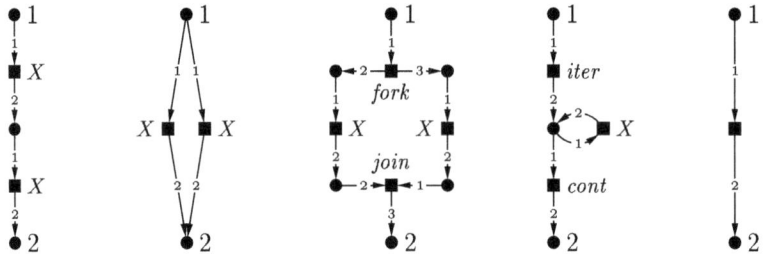

Figure 6.18: The right-hand sides of the HR productions

6.14 Construction (workflow net grammar)
Let $HG_{WF} = (N, T, P, Z)$ be a hyperedge-rewriting grammar with a unique nonterminal symbol X, terminal symbols $*$, *fork*, *join*, *iter*, and *cont*, five productions with left-hand side X and as right-hand side one of the hypergraphs with external nodes of Figure 6.18, and as axiom:

Then on the basis of Translation 2, the axiom corresponds to N_{basic} and the first four productions to the transition refinement rules of Figure 6.16, where the direction functions of the five labels are indicated by the directions on the hyperedge tentacles (and, as only net structures—without tokens—are considered, $*$ is the only node label).

Obviously, any terminal hypergraph of HG_{WF} can be interpreted as a workflow net derived from N_{basic} by using the same translation as in the construction. Moreover, as HG_{WF} is context-free, the usual notion of derivation trees can be exploited for a simple method to reengineer a model of a workflow process, by cutting off some branches of a particular model's derivation tree and plugging in some other branches.

6.5 Concluding Remarks

The considerations of the present chapter illustrate that context-free hypergraph rewriting provides a formal basis for net refinement. While some refinement techniques can be formulated suitably as context-free node rewriting in graphs or hyperedge rewriting in hypergraphs, at least the technique presented in [GG90] requires the flexibility of context-free node rewriting in hypergraphs, namely the multiplication and transformation of embedding hyperedges. It may be interesting to study systematically further net refinement techniques from this perspective.

A new trend in Petri net refinement is the implicit use of grammars with refinement rules tailored to suit the needs of a particular application area. The workflow net refinements of [vdA97], for instance, allow to construct and modify models of workflow processes. A further example is supplied in [Zub98] where Petri nets are used to model schedules for manufacturing cells; the focus here lies on generating all Petri net models of schedules for cells with n machines ($n \in \mathbb{N}$). In this reference, however, the net refinements are not fully specified, and it seems doubtful that the proposed rules do actually allow to obtain all required nets. Nevertheless, the use of 'refinement grammars' suggests that formal language theory in the guise of context-free hypergraph grammars may prove useful for particular Petri net applications.

Finally, it should be noted that general (hyper)graph transformation allows to model the dynamic aspects of Petri nets, i.e. the firing of transitions, too [Kre81].

7

Category-theoretical Treatment

While (hyper)edge rewriting has grown on the solid category-theoretical founda-
tion of the double-pushout approach [Ehr79], such a framework for node rewriting
has been proposed but recently with Bauderon and Jacquet's pullback approach
[Bau95a, Bau96, Jac99, BJ01a]. The approach can express node rewriting in graphs
[Bau95a] as well as in hypergraphs [BJ01b, JK00], provides a model for the generation of infinite graphs [Bau96], and can deal with the parallelisation of node
rewriting [Bau95b]. Moreover, the construction of a pullback of simple graph morphisms is easy to compute automatically, so pullback rewriting may be a suitable
basis for a general implementation of node rewriting. Finally, the ideas underlying
the pullback approach have proved to be stable enough to be transferred to an
application such as refinement in Petri nets [Kle98].

   The aim of this chapter is to show how node rewriting in hypergraphs on the one
hand and net refinement on the other fit in the category-theoretical framework of
pullback rewriting. It is organised as follows: Section 7.1 recalls some basic concepts
from category theory. The category of graphs and derived comma categories are
defined in Section 7.2. The general principles underlying pullback rewriting are
presented in Section 7.3. Section 7.4 contains the translation of hNCE rewriting
into pullback rewriting, and Section 7.5 that of transition refinement à la [GG90].
Finally, some concluding remarks can be found in Section 7.6.

7.1   Categories, Products, and Pullbacks

This section recalls some basic category-theoretical concepts. It is oriented at
[AL91] and [HS79].

A *category* is a tuple $Cat = (Ob_{Cat}, Mor_{Cat}, dom, cod, \circ)$ where Ob_{Cat} is a class of *objects*, Mor_{Cat} is a class of *morphisms* (or *arrows*), *dom* and *cod* are functions assigning to each morphism f a *domain* $dom(f) \in Ob_{Cat}$ and a *codomain* $cod(f) \in Ob_{Cat}$, and $\circ$ is a function assigning to each pair of morphisms (f, g) with $dom(f) = cod(g)$ its *composition* $f \circ g \in Mor_{Cat}$ with $dom(f \circ g) = dom(g)$ and $cod(f \circ g) = cod(f)$, such that the following conditions hold:

- ASSOCIATIVITY: For all morphisms f, g, h with $dom(f) = cod(g)$ and $dom(g) = cod(h)$ we have $(f \circ g) \circ h = f \circ (g \circ h)$.

- IDENTITY EXISTENCE: For each object $A \in Ob_{Cat}$ there exists a morphism $id_A \in Mor_{Cat}$ such that $dom(id_A) = A = cod(id_A)$ and for all morphisms f, g with $dom(f) = A = cod(g)$ we have $f \circ id_A = f$ and $id_A \circ g = g$.

- SMALLNESS OF MORPHISM CLASS: For each pair of objects (A, B) the class $\hom_{Cat}(A, B) = \{f \in Mor_{Cat} \mid dom(f) = A \text{ and } cod(f) = B\}$ is a set.

A morphism f with domain A and codomain B may be written $f \colon A \to B$ or $A - f \to B$.

An object T is a *terminal* object if for all objects A there is exactly one morphism in $\hom_{Cat}(A, T)$.

A *product* of a pair (A, B) of objects is a triple (P, π_A, π_B) where P is an object (called the *product* object) and $\pi_a \colon P \to A$, $\pi_B \colon P \to B$ are morphisms (called *projections*) such that if C is an object and $f \colon C \to A$, $g \colon C \to B$ are morphisms, then there exists a unique morphism $\langle f, g \rangle \colon C \to P$ with $f = \pi_A \circ \langle f, g \rangle$ and $g = \pi_B \circ \langle f, g \rangle$. For each pair of objects, if the product (object) exists then it is unique up to isomorphism.

A *pullback*, also called *fibred product*, of a pair $B_1 - f_1 \to C \leftarrow f_2 - B_2$ of morphisms with common codomain is a triple (A, g_1, g_2) where A is an object (called the *pullback* object) and $B_1 \leftarrow g_1 - A - g_2 \to B_2$ are morphisms (called *projections*) such that $f_1 \circ g_1 = f_2 \circ g_2$, and if D is an object and $B_1 \leftarrow h_1 - D - h_2 \to B_2$ are morphisms with $f_1 \circ h_1 = f_2 \circ h_2$, then there is a unique morphism $D - f \to A$ with $h_1 = g_1 \circ h$ and $h_2 = g_2 \circ h$. For each pair of morphisms, if the pullback (object) exists then it is unique up to isomorphism.

The diagrams in Figures 7.1–7.3 illustrate the constructions of a terminal object, a product, and a pullback. The morphism drawn as a dashed arrow is the one of which unique existence is required. In most textbooks on category theory, the diagrams for a product and a pullback have a different layout (and due to its simplicity, the diagram for a terminal object is not given at all). In particular, the pullback square, i.e. the square formed by the original morphisms and the morphisms of the pullback, is usually drawn in such a way that the pullback object is placed in the top left-hand corner. The purpose of the orientation chosen here

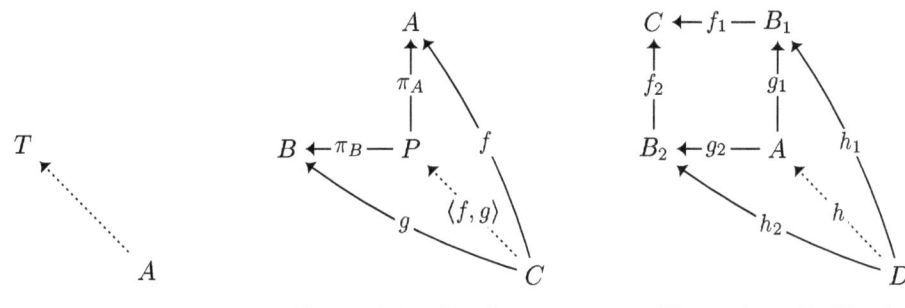

Figure 7.1: Terminal object T

Figure 7.2: Product (P, π_A, π_B) of (A, B)

Figure 7.3: Pullback (A, g_1, g_2) of (f_1, f_2)

is to support an intuitive understanding of the development from the bottom left-hand object to the pullback object as a rewriting step, see Section 7.3.

Terminal objects, products, and pullbacks are examples for a special type of categorical constructions called *limits*, and a category in which all limits exist is called *complete*.

The following relationships hold for terminal objects, products, and pullbacks (see e.g. Exercise 18I and Proposition 21.5 in [HS79]).

7.1 Fact
Let Cat be a category with terminal object T.

1. *For every object A in Cat, the product of T and A exists and the product object is isomorphic to A.*

2. *The triple (A, g_1, g_2) is the pullback of two morphisms with common codomain T if and only if (A, g_1, g_2) is the product of $(cod(g_1), cod(g_2))$.*

Let *Cat* be a category and S an object in *Cat*. The *slice category* or *comma category* of *Cat* over S is the category $\langle Cat \downarrow S \rangle$ whose objects are all morphisms of *Cat* with codomain S, and whose morphisms from $A - f \rightarrow S$ to $B - f' \rightarrow S$ are all morphisms $A - g \rightarrow B$ in *Cat* such that $f = f' \circ g$, i.e. such that the following triangle commutes:

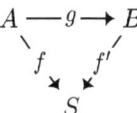

7.2 Fact (properties of comma categories)
Let $\langle Cat \downarrow S \rangle$ be the comma category of Cat over S.

1. *The terminal object of $\langle Cat \downarrow S \rangle$ is $id_S \colon S \rightarrow S$.*

2. If Cat is complete, then so is $\langle \mathit{Cat} \downarrow S \rangle$.

3. If S is the terminal object of Cat, then $\langle \mathit{Cat} \downarrow S \rangle$ and Cat are isomorphic categories.

7.2   A Category of Graphs

In its basic version, pullback rewriting is based on undirected, unlabelled, simple graphs; cf. [Bau95a, BJ01a]. In this section, the associated category of graphs and graph morphisms is recalled.

A(n undirected, simple) *graph* is a tuple $G = (V_G, E_G)$ consisting of a set V_G of *vertices* and a set $E_G \subseteq V_G \times V_G$ of *edges* with $\langle u, v \rangle \in E_G$ if and only if $\langle v, u \rangle \in E_G$. For all $u, v \in V_G$, we consider the edges $\langle u, v \rangle$ and $\langle v, u \rangle$ to be the same. In pictures of a graph, a vertex is drawn as a circle $\circ$ and an edge between two vertices as a line $\circ\!\!-\!\!\circ$. An edge $\langle v, v \rangle \in E_G$ is a *loop*, and a vertex $v \in V_G$ which has a loop is said to be *reflexive*. A vertex $u \in V_G$ is a *neighbour* of a vertex $v \in V_G$ if there is an edge $\langle u, v \rangle$ in G. For a graph $G = (V_G, E_G)$ and a set $V \subseteq V_G$, the *subgraph of G induced by V* is $G|_V = (V, E)$ with $E = E_G \cap (V \times V)$.

Let G, G' be graphs. A *graph morphism* $f \colon G \to G'$ from G to G' is a pair of mappings $f = \langle f_V, f_E \rangle$ with $f_V \colon V_G \to V_{G'}$ and $f_E \colon E_G \to E_{G'}$ such that $f_E(\langle u, v \rangle) = \langle f_V(u), f_V(v) \rangle$ for all $\langle u, v \rangle \in E_G$. We also write $f(x)$ for $f_V(x)$ or $f_E(x)$. A graph morphism $f = \langle f_V, f_E \rangle$ is an *isomorphism* if both f_V and f_E are bijective. Note that a graph morphism $f \colon G \to G'$ is completely specified with the images of the vertices of G.

7.3 Fact (category of graphs)
Graphs and graph morphisms form a category, which is denoted by $\mathcal{G}$. This category is complete. The terminal object of $\mathcal{G}$ is the single reflexive vertex $\circ\!\!\circlearrowleft$. The pullback of two graph morphisms $f \colon B \to A$, $g \colon C \to A$ consists of a graph D and two graph morphisms $f' \colon D \to B$, $g' \colon D \to C$, where $V_D = \{(u, v) \in V_B \times V_C \mid f(u) = g(v)\}$, $E_D = \{\langle (u, v), (u', v') \rangle \in V_D \times V_D \mid \langle u, u' \rangle \in E_B \text{ and } \langle v, v' \rangle \in E_C\}$, and for all $(u, v) \in V_D$, $f'((u, v)) = u$ and $g'((u, v)) = v$.

Figure 7.4 shows an example of a pullback in $\mathcal{G}$. The morphisms are indicated by the shades of grey of the nodes and by relative spatial arrangement; f maps, e.g., the two white vertices of B to the white vertex of A, and f' maps the upper white vertex of D to the upper white vertex of B.

Let S be a graph. A *graph structured by S* or *S-graph* is a graph morphism $\bar{G} \colon G \to S$, i.e. an object in the slice category $\langle \mathcal{G} \downarrow S \rangle$. In order to draw structured graphs, we will always choose distinct shapes for the vertices of the structuring

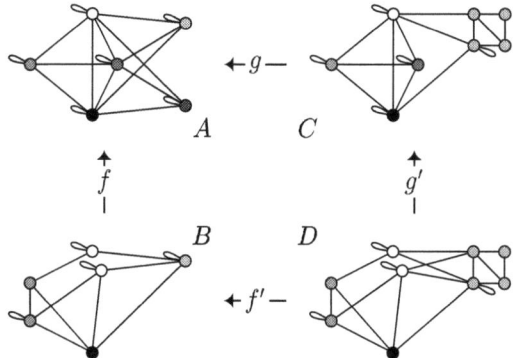

Figure 7.4: A pullback of graph morphisms

graph S and represent the morphism from a graph to S by drawing a vertex of
the graph in the shape of the vertex on which it is mapped by the morphism; see
Figure 7.5 for an example. Moreover, if a vertex v of S is assigned some shape $\diamond$,
then a vertex of an S-graph which is mapped on v is also called a $\diamond$-vertex.

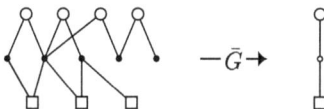

Figure 7.5: Drawing a structured graph

The pullback $(\bar{D}, f', g')$ of two S-graph morphisms f and g always exists by
Fact 7.3 and 7.2. It is constructed just as the pullback (D, f', g') of the associated
graph morphisms f and g, with the pullback object $\bar{D} \colon D \to S$ mapping a vertex
of D on a vertex of S such that f' and g' are S-graph morphisms, see Figure 7.6.

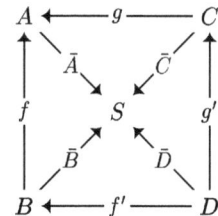

Figure 7.6: Pullback of S-graph morphisms f and g

7.3   Pullback Rewriting

The general idea for pullback rewriting is illustrated in Figure 7.7. Let *Cat* be a category with a terminal object T. Suppose that in an object B, an occurrence of T is to be replaced with some object R. Then B can be divided into this occurrence of T (the *unknown* part), its farther context B^- which is not altered by the replacement (the *context* part), and a part of B which links the occurrence of T with B^- (the *interface* part). This division of B can be expressed by a morphism f to an object A which is divided in the same way, with one copy of T as context part and many copies of T in the unknown part. Moreover, a second morphism g from an object C (containing R) to the same A also expresses a division of C in unknown, context, and interface part, with the unknown part consisting of R and the context part being a copy of T.

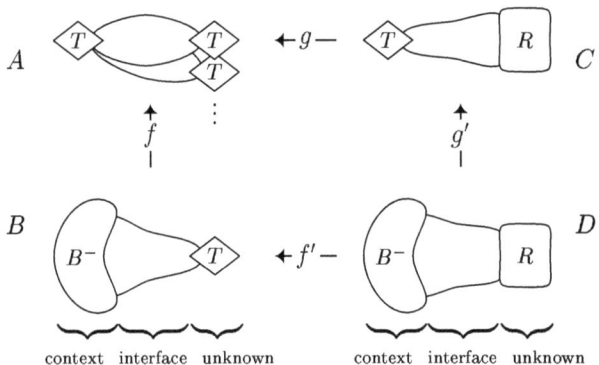

Figure 7.7: Pullback rewriting in general

If the morphisms f and g map the unknown part of B resp. C on the same copy of T in the unknown part of A, then constructing their pullback can be interpreted as replacing the occurrence of T in B with R. More precisely, if D is the pullback object, it contains an isomorphic copy of the context part B^- of B since the context parts of A and C are terminal objects, and an isomorphic copy of the unknown part R of C since the unknown parts of A and B are terminal objects (cf. Fact 7.1). Thus, one can say that B derives to D by applying the *rule* g to the *unknown morphism* f. Read from left to right, the bottom line of the pullback square then describes a rewriting step obtained by an application of the rule drawn in the top line of the square.

In this framework, the concept of nonterminal label resp. left-hand side of a rule is expressed by the copy of T in A on which f resp. g map the unknown part of B resp. C. Finally, note that the pullback construction allows to 'multiply' items of

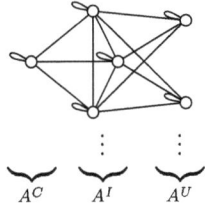

Figure 7.8: The generic alphabet graph

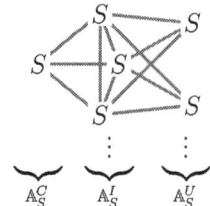

Figure 7.9: The structured alphabet graph (schematic)

the interface part of B with items of the interface part of C if all these items are mapped on the same item in the interface part of A.

Below, pullback rewriting is defined directly for the case of graphs structured by some fixed graph S. First, the general form of the common codomain of unknown and rule morphisms is determined. As this graph plays a role very similar to that of an alphabet of labels for labelled graphs, it is called the alphabet graph. Moreover, there is a generic form of the alphabet graph, which is then altered with the structuring graph S to yield the structured alphabet graph for rewriting in $\langle \mathcal{G} \downarrow S \rangle$ ([BJ01a]; see also [Jac99, Section 6.3.2]).

7.4 Definition (alphabet graph)
The *generic alphabet graph* A consists of

- a countably infinite set A^U of reflexive vertices (we will speak of the ith unknown vertex in A),

- a countably infinite set A^I of reflexive vertices (we will speak of the jth interface vertex in A),

- a set A^C of one reflexive vertex (we will speak of the context vertex in A), and

- all possible edges between A^U and A^I, between A^I and A^C, and between vertices of A^I.

Let S be a graph. The *structured alphabet graph* in $\langle \mathcal{G} \downarrow S \rangle$ is the second projection $\bar{\mathbb{A}}_S : \mathbb{A}_S \to S$ of the product $(\mathbb{A}_S, \pi_A, \bar{\mathbb{A}}_S)$ of A and S. The part of $\mathbb{A}_S$ which π_A projects on the ith unknown vertex is called the *i-copy of S in* $\bar{\mathbb{A}}_S^U$. Whenever S is clear from the context, we will omit the subscript S.

Figure 7.8 illustrates the construction of the generic alphabet graph. Note that the graph A in Figure 7.4 is a finite subgraph of the generic alphabet graph with two unknown vertices drawn on the right, three interface vertices drawn in the middle, and the context vertex drawn on the left. In concrete examples such as this, we will always use a 'sufficiently large' subgraph of the alphabet graph and call it A, too.

Figure 7.9 illustrates the construction of the structured alphabet graph. Each of the grey edges stands for all edges between vertices of the two instances of S which comply with the structure of S. Taking as structuring graph e.g. the three-node string graph ∘↗□ as in Figure 7.5, the structured alphabet contains (among others) edges between the •-vertex of $\bar{A}^C$ and all ∘- and □-vertices of $\bar{A}^I$.

The following definitions of unknown and rule morphisms are meant to capture a very general idea of these notions. In particular, they leave open to further specification what constitutes the interface part of these morphisms.

7.5 Definition (unknown morphism)

Let $\bar{G}\colon G \to S$ be an S-graph and X an occurrence (i.e. an isomorphic copy) of S in G. An *unknown on* X is an S-graph morphism $\varphi_X\colon \bar{G} \to \bar{A}$ mapping X to one copy of S in $\bar{A}^U$ (φ_X is said to have the *type i* if $\varphi_X(X)$ is the i-copy of S), the vertices of the 'nearer context' of X in G to $\bar{A}^I$, and all other vertices of G (the 'farther context' of X) to $\bar{A}^C$.

7.6 Definition (rule morphism)

A *rule* is an S-graph morphism $r\colon \bar{R} \to \bar{A}$ (where $\bar{R}\colon R \to S$) which maps all vertices of R in $r^{-1}(\bar{A}^U)$ to the same copy of S in $\bar{A}^U$ (r is said to have the *type i* if this is the i-copy of S) and whose restriction to the subgraph of R induced by $r^{-1}(\bar{A}^C)$ and the subgraph $\bar{A}^C$ of $\bar{A}$ is an isomorphism.

7.7 Definition (application of a rule to an unknown)

Let $\varphi_X\colon \bar{G} \to \bar{A}$ be an unknown and $r\colon \bar{R} \to \bar{A}$ a rule of the same type. The *application* of r to φ_X is obtained by computing the pullback of φ_X and r, and we write $\bar{G} \to_{(\varphi_X, r)} \bar{G}'$, where $\bar{G}'$ is the pullback object.

The pullback of Figure 7.4 can be interpreted as a vertex rewriting step in the category $\mathcal{G}$ (or $\langle \mathcal{G} \downarrow \infty \rangle$; recall that these categories are essentially the same), where the rightmost, reflexive vertex v of B is replaced with the graph

$$C^- = \boxtimes$$

such that each reflexive (white) neighbour of this node is connected to each of the left vertices in C^-, and each (black) neighbour which is not reflexive is connected to the reflexive vertex of C^-. The graph morphism f is an unknown on v in B which maps exactly the neighbours of v to A^I. The graph morphism g is a rule which maps C^-—the subgraph of C induced by the vertices in $g^{-1}(A^U)$—to one unknown vertex, and which is an isomorphism on the subgraph of C induced by all vertices not in C^- and the subgraph of A induced by A^I and A^U. The pullback of f and g is the rule application $B \rightarrow_{(f,g)} D$. Its construction reproduces C^- because its image $g(C^-)$ in A^U is isomorphic to the inverse image $f^{-1}(g(C^-))$ of this image under f (both consist of a single reflexive vertex, the terminal object of $\mathcal{G}$). Similarly, the isomorphy between the context parts of A and C which is required for a rule guarantees that B^- is reproduced. Finally, all neighbours of v in B are reproduced because g is also an isomorphism on the interface vertices of A and C.

Note that f distinguishes between neighbours of v which are going to be treated differently during the replacement of v by mapping them to distinct interface vertices of A. On the other hand, g specifies the way in which C^- is linked to (white resp. black) neighbours of the vertex to be rewritten. In other vertex-rewriting approaches, e.g. NLC or NCE grammars, these distinctions are achieved by assigning distinct labels to the vertices.

Finally, it must be mentioned that the definition of unknown and rule morphisms as given above is at the same time more general and more restrictive than the definition given in [BJ01a, Section 3]. There, the interface part of the domain of an unknown resp. of a rule is required to consist only of neighbours of a vertex in the respective unknown part, while the unknown part of an unknown resp. the context part of a rule need not be a full copy of S. As witnessed by the following translation of hNCE rewriting into the pullback approach, the former restriction is too severe for our purposes, while the latter generality is not needed here. All in all, this led to the definitions as given here.

7.4   Simulation of hNCE Rewriting

In [Bau96], NLC node rewriting (and, in fact, NCE rewriting) in graphs is translated into terms of pullback rewriting. A notion of node rewriting in hypergraphs with the pullback approach is presented in [BJ01b], where hypergraphs are seen as bipartite graphs or, more precisely, as graphs structured by the two-vertex loop-free graph $\circ\!\!-\!\!\square$. Coding hypergraphs as this kind of structured graphs is, however, not rich enough to capture *directed* hypergraphs which have hyperedges with a sequence of tentacles such as the hypergraphs of [HK87] or [Kle96]. As proposed in [BJ01a], other graphs than the two-node loop-free one can be used to construct graph-based objects with a more involved structure. In fact, the linear three-vertex loop-free graph $\circ\!\!\overset{\wedge}{}\!\!\square$ allows to materialise tentacles in the form of $\bullet$-vertices. On this basis, hNCE node rewriting in directed hypergraphs can be translated into the pullback approach [JK00] while staying in the general framework as presented in the previous section. The focus in this section lies on the encoding of a derivation step, i.e. given a hypergraph with a nonterminal node and an hNCE production which can be applied to that node, the corresponding unknown and rule morphisms are constructed and shown to be correct.

For the translation into pullback rewriting, we consider hNCE rewriting in multiple hypergraphs because the constructions will be less complicated this way. By identifying parallel hyperedges into one, the original notion of hNCE rewriting in simple hypergraphs is obtained.

7.8 Definition (multiple hypergraph with embedding)
Let $EX^\circ = (\Sigma \times (\Sigma \uplus \{\Diamond\})^*)^* \setminus (\Sigma \times \Sigma^*)$ be the (countably infinite) set of local existence parts over Σ, $CR_M = \Sigma \times (\mathbb{N}_+ \uplus V_M)^*$ the set of creation parts for a multiple hypergraph M, and

$$CI_M^\circ = \{(ex/cr) \in EX^\circ \times CR_M \mid \forall j \in [rank(cr)] \colon cr[j] \in \mathbb{N}_+ \Rightarrow$$
$$cr[j] \in [rank(ex)] \wedge ex[cr[j]] \in \Sigma\}$$

the set of local connection instructions over Σ for M. A *multiple hypergraph with embedding* over Σ is a pair (M, C) where M is a multiple hypergraph over Σ and the connection relation $C \subseteq CI_M^\circ$ is a finite set of local connection instructions.

7.9 Definition (hNCE rewriting in multiple hypergraphs)
Let (M_1, C_1), (M_2, C_2) be two disjoint multiple hypergraphs with embedding and v a node in M_1. Then $(M_1, C_1)[v/(M_2, C_2)]$ is the multiple hypergraph with embedding (M_3, C_3) which is constructed just as in the case of simple hypergraphs (cf. Definition 2.6), with the exception of the hyperedges: if $M_3 = (V_3, E_3, lab_3, att_3)$, then

$$E_3 = (E_1 \setminus \{e \mid v \in vset_1(e)\}) \cup E_2 \cup$$
$$\{(e, (ex/cr)) \in E_1 \times C_2 \mid lab_1(e) = lab(ex),\ rank_1(e) = rank(ex),$$
$$\forall i \in \{rank_1(e)\} \colon (att_1(e, i) = v \wedge ex[i] = \Diamond)$$
$$\vee (att_1(e, i) \neq v \wedge ex[i] = lab_1(att_1(e, i)))\}$$

with

- $lab_3(e) = lab_1(e)$ and $att_3(e) = att_1(e)$ if $e \in E_1 \setminus \{e' \mid v \in vset_1(e')\}$,

- $lab_3(e) = lab_2(e)$ and $att_3(e) = att_2(e)$ if $e \in E_2$, and

- if $e = (e', (ex/cr)) \in E_1 \times C_2$:
  $lab_3(e) = lab(cr)$, $rank_3(e) = rank(cr)$, and for all $j \in \{rank(cr)\}$:

$$att_3(e, j) = \begin{cases} cr[j] & \text{if } cr[j] \in V_2, \text{ and} \\ att_1(e', cr[j]) & \text{if } cr[j] \in [rank(ex)]. \end{cases}$$

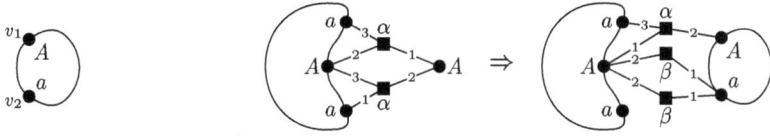

Figure 7.10: R Figure 7.11: Application of $A ::= (R, C)$ to a node

The main interest of this section lies in showing how an hNCE rewriting step can be translated into a pullback rewriting step. We will therefore assume that the alphabet Σ is partitioned into an alphabet N of nonterminal labels and an alphabet T of terminal labels, and that productions and derivation steps are defined as usual.

7.10 Example (hNCE rewriting in multiple hypergraphs)

Let $A \in N$ and $a \in T$; $\alpha, \beta \in \Sigma$ may be either terminal or nonterminal. Consider a production $A ::= (R, C)$ where R is some multiple hypergraph with (at least) nodes v_1 and v_2 as sketched in Figure 7.10, and $C = \{(ex_1/cr_1), (ex_1/cr_2), (ex_2/cr_3)\}$ with:

$$ex_1 = (\alpha, \Diamond Aa) \qquad cr_1 = (\alpha, 2v_13)$$
$$cr_2 = (\beta, v_22)$$
$$ex_2 = (\alpha, a\Diamond A) \qquad cr_3 = (\beta, v_23)$$

Applying this production to the rightmost node of the left hypergraph in Figure 7.11 yields the hypergraph on the right, where the upper α-labelled hyperedge gives rise to two embedding hyperedges labelled α resp. β, and the lower α-labelled hyperedge results in one β-labelled embedding hyperedge. Note that the two β-labelled embedding hyperedges are parallel.   ∎

In order to use the general framework for pullback rewriting and to obtain a suitable category for the representation of multiple hypergraphs, we transform multiple hypergraphs into a type of structured graphs we call *structured hypergraphs*, which are subsequently *expanded* to code the labelling, too. In the following, we present the ideas for these constructions rather than a full formalisation.

7.11 Construction (expanded structured hypergraph)

A directed labelled hypergraph is transformed into an expanded structured hypergraph in three steps, as illustrated in Figure 7.12.

STEP 1: The basic idea to represent directed hypergraphs as structured hypergraphs is to code nodes (∘) and hyperedges (□) as two sorts of vertices, and introduce tentacles (•) as a third sort to describe the attachment of a node to a hyperedge. So, let us fix S as the graph ∘‾□, i.e. as the graph with vertices ∘, •, □ and edges $\langle \circ, \bullet \rangle$, $\langle \bullet, \square \rangle$, $\langle \square, \bullet \rangle$, $\langle \bullet, \circ \rangle$. A vertex which an S-graph maps on ∘ is called a ∘-vertex, and analogously for □-vertices and •-vertices.

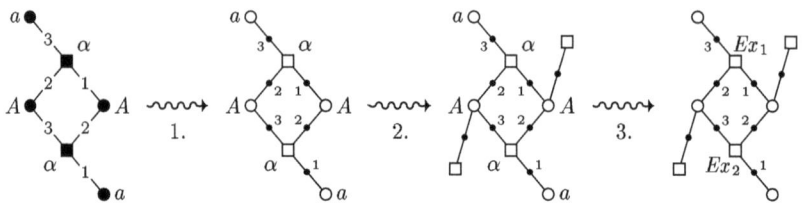

Figure 7.12: Transforming a labelled hypergraph into an expanded structured hypergraph (A is nonterminal, and $ex_1 = (\alpha, AAa)$, $ex_2 = (\alpha, aAA)$, with corresponding expansions Ex_1, Ex_2)

Note that not all objects of $\langle \mathcal{G} \downarrow S \rangle$ are good candidates to represent directed hypergraphs: As a •-vertex represents a tentacle, it must be adjacent to exactly one ○-vertex and one □-vertex.

STEP 2: The pullback approach rewrites a whole pattern—a copy of the terminal object—rather than just a vertex. Thus, in the category of graphs, the vertex to be rewritten has to be reflexive, and in the category of S-graphs, the pattern to be rewritten is a copy of S. Therefore, every nonterminally labelled node must be completed to such a copy by adding new private •- and □-vertices to the ○-vertex representing it.

STEP 3: Finally, to collect the information the hNCE approach needs in order to classify the hyperedges, we expand the label of a hyperedge with the labels of the nodes incident to that hyperedge.

For an existence part $ex \in EX^{\circ}$, let $\bullet ex = \{i \in [rank(ex)] \mid ex[i] \neq \Diamond\}$ and $ex\bullet = \{i \in [rank(ex)] \mid ex[i] = \Diamond\}$. Then $Ex = (\bullet ex, ex, ex\bullet)$ is the *expansion triple* associated with ex. Analogously, $Cr = (\bullet cr, cr, cr\bullet)$ is the expansion triple associated with a creation part $cr \in CR_M$, where $\bullet cr = \{i \in [rank(cr)] \mid cr[i] \in \mathbb{N}_+\}$ and $cr\bullet = \{i \in [rank(cr)] \mid cr[i] \in V_M\}$.

We *expand* the label $lab(e)$ of a hyperedge e into the expansion triple Ex_e associated with the existence part $ex_e = (lab(e), lab(att(e)))$. For example, the expansion triple Ex_1 associated with $ex_1 = (\alpha, AAa)$ of Figure 7.12 has $\bullet ex_1 = \{1, 2, 3\}$ and $ex_1\bullet = \emptyset$. (The component $ex\bullet$ will be used when coding a rewriting step.)

TO CONCLUDE. A *structured hypergraph* is a graph morphism $\bar{H} : H \to S$ such that each •-vertex u of V_H is adjacent to exactly one ○-vertex and one □-vertex and each nonterminal node is completed to a copy of S by a private tentacle and a private hyperedge. An *expanded structured hypergraph* is a structured hypergraph in which all hyperedge labels are expanded.

Figure 7.13: Step-by-step construction of the structured alphabet graph

7.12 Construction (structured alphabet graph)

The *structured alphabet graph* associated with Σ is the S-graph $\bar{\mathbb{A}}\colon \mathbb{A} \to S$ constructed as follows, with Figure 7.13 providing an illustration:

UNKNOWN PART: Take a countably infinite set of isomorphic copies of S (we will speak of the *k-copy of S* in $\bar{\mathbb{A}}^U$, for $k \in \mathbb{N}_+$).

INTERFACE PART: For each element $ex \in EX^\circ$, take an S-graph $\lfloor Ex \rfloor$ which consists of one $\square$-vertex standing for $Ex = (\bullet Ex, Ex, Ex\bullet)$ and $rank(ex)$ many $\bullet$-vertices linked to it by an edge, and put an edge between each $\circ$-vertex of $\bar{\mathbb{A}}^U$ and each $\bullet$-vertex in $ex\bullet$.

CONTEXT PART: Take an isomorphic copy $\bar{\mathbb{A}}^C$ of S and link its $\circ$-vertex to each $\bullet$-vertex in every $\bullet ex$.

The structured alphabet graph $\bar{\mathbb{A}}$ as constructed above is a subgraph of the general form $\bar{\mathbb{A}}_S\colon \mathbb{A}_S \to S$ from Definition 7.4. In particular, $\bar{\mathbb{A}}^I$ does not contain the $\circ$-vertices of $\bar{\mathbb{A}}^I$, and the edges in $\bar{\mathbb{A}}$ are a proper subset of the edges in $\bar{\mathbb{A}}_S$. Thus, the construction of $\bar{\mathbb{A}}$ may also be understood as distinguishing those parts of $\bar{\mathbb{A}}_S$ which are needed to encode hNCE rewriting.

Note, moreover, that $\bar{\mathbb{A}}$ is not a structured hypergraph because some $\bullet$-vertices in the interface part are incident to more than one $\circ$-vertex.

7.13 Construction (unknown)

Let v be a 'nonterminal' $\circ$-vertex of an expanded structured hypergraph $\bar{G}$ and X the copy of S defined by v, i.e. v together with its private $\square$- and $\bullet$-vertices. An *unknown* on X of type k is a morphism $\varphi_X\colon \bar{G} \to \bar{\mathbb{A}}$ constructed as follows (see the left half of Figure 7.14 for an example of such an unknown):

UNKNOWN PART: Map X to the k-copy of S in $\bar{\mathbb{A}}^U$.

INTERFACE PART: For each hyperedge e adjacent to v, first transform its expansion triple $Ex_e = (\bullet ex_e, ex_e, ex_e\bullet)$ to Ex'_e as follows: for all $i \in [rank(ex_e)]$, if $att(e, i) = v$ then move i from $\bullet ex_e$ to $ex_e\bullet$ and replace $ex_e[i]$ by the symbol $\Diamond$. Then map e to the $\square$-vertex of $\lfloor Ex'_e \rfloor$ in $\bar{\mathbb{A}}^I$, and map its $\bullet$-vertices to their corresponding $\bullet$-vertices in $\lfloor Ex'_e \rfloor$.

In Figure 7.12 (see also Figure 7.14, bottom left), choose for example the right A-labelled node v. Then the expansion triples corresponding to the hyperedges

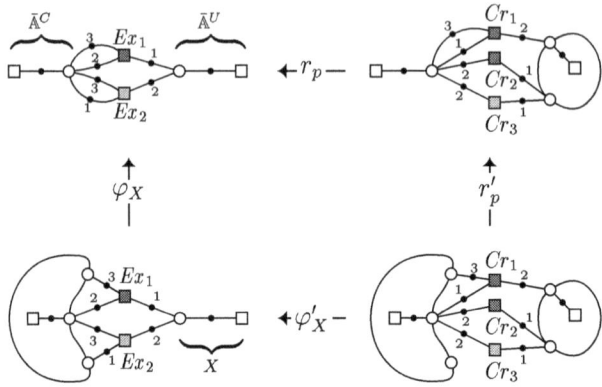

Figure 7.14: Application of r_p to φ_X (cf. Figure 7.11)

with respect to x as nonterminal are $Ex_1 = (\{2,3\}, (\alpha, \Diamond Aa), \{1\})$ and $Ex_2 = (\{1,3\}, (\alpha, a\Diamond A), \{2\})$.

CONTEXT PART: Map the rest of G to $\bar{\mathbb{A}}^C$.

7.14 Construction (rule)

Let $p = (A ::= (R, C))$ be a production. A *rule* of type k associated with p is a morphism $r_p \colon \bar{R}_p \to \bar{\mathbb{A}}$ constructed as follows[1] (see the upper half of Figure 7.14 for an example of such a rule):

UNKNOWN PART: Take the structured hypergraph $\bar{R}^U \colon R^U \to S$ corresponding to R and let r_p map it to the k-copy of S in $\bar{\mathbb{A}}^U$.

INTERFACE PART: For each $(ex/cr) \in C$, take an S-graph $\lfloor Cr \rfloor$ as described for Ex and link each tentacle i in $cr\bullet$ to the $\circ$-vertex $cr[i] \in V_R$. Let r_p map the $\square$-vertex of $\lfloor Cr \rfloor$ to the $\square$-vertex of $\lfloor Ex \rfloor$ in $\bar{\mathbb{A}}^I$, each tentacle $i \in \bullet cr$ to the tentacle $cr[i] \in \bullet ex$, and each tentacle in $cr\bullet$ to one of the tentacles in $ex\bullet$.[2] Finally, expand the label of the $\square$-vertex of $\lfloor Cr \rfloor$ into the expansion triple $Ex(cr)$ associated with the existence part $ex(cr) = (\alpha, x_1 \ldots x_{rank(cr)})$ where $x_i = lab_R(cr[i])$ if $cr[i] \in V_R$ and $x_i = ex[cr[i]]$ if $cr[i] \in \mathbb{N}_+$.

CONTEXT PART: Take one copy of S, link the $\circ$-vertex to each $\bullet$-vertex in every $\bullet cr$, and let r_p map this S-copy to $\bar{\mathbb{A}}^C$.

The resulting structured hypergraph is $\bar{R}_p$, which is expanded by taking the expansion of $\bar{R}^U$ and associating with each $\square$-vertex introduced for a connection instruction $(ex/cr) \in C$ the expansion triple $Ex(cr)$.

---

[1] In general, r_p is not unique –

[2] – because there may be more than one tentacle in $ex\bullet$. Note, moreover, that $ex\bullet$ is not empty because $ex \in EX^\circ$ is local.

7.15 Definition (application of a rule to an unknown)

Let $\varphi_X \colon \bar{G} \to \bar{A}$ be an unknown and $r_p \colon \bar{R}_p \to \bar{A}$ a rule of the same type. The *application* of r_p to φ_X is obtained by computing the pullback of φ_X and r_p, and we write $\bar{G} \to_{(\varphi_X, r_p)} \bar{H}$, where $\bar{H}$ is the pullback object. Moreover, $\bar{H}$ is expanded by taking, for its context part, the expansion from $\bar{G}$, and for its interface and unknown part, the expansion from $\bar{R}_p$.

7.16 Example (translation into pullback rewriting)

Figure 7.14 shows the translation of the derivation step from Example 7.10 into the pullback approach. The morphism on the left is the unknown corresponding to the nonterminal node, and the morphism at the top the rule corresponding to the production. Read from left to right, the lower half shows the derivation. Note that the expansion triples are only added to indicate the references. ∎

7.17 Lemma

The pullback object of a rule application is a structured hypergraph.

Proof. Consider the application of a rule $r_p \colon \bar{R}_p \to \bar{A}$ to an unknown $\varphi_X \colon \bar{G} \to \bar{A}$ on X, where $(\bar{H} \colon H \to S,\ \varphi'_X \colon \bar{H} \to \bar{G},\ r'_p \colon \bar{H} \to \bar{R}_p)$ is the pullback of (φ_X, r_p) in the category $\langle \mathcal{G} \downarrow S \rangle$. Then $\bar{H}$ is an S-graph.

The context part of $\bar{H}$ (which is mapped to $\bar{A}^C$ by $\varphi_X \circ \varphi'_X$) is an isomorphic copy of the context part of $\bar{G}$ (which is mapped to $\bar{A}^C$ by φ_X). Furthermore, the unknown part of $\bar{H}$ (which is mapped to $\bar{A}^U$ by $r_p \circ r'_p$) is an isomorphic copy of the unknown part $\bar{R}^U$ of $\bar{R}_p$ (which is mapped to $\bar{A}^U$ by r_p). Consequently, the context and unknown parts of $\bar{H}$ are structured hypergraphs.

Now consider a •-vertex x of $\bar{H}$ which is mapped to a •-vertex x' in $\bar{A}^I$ (by $\varphi_X \circ \varphi'_X$ as well as $r_p \circ r'_p$), and let x_1 denote the vertex $\varphi'_X(x) \in V_G$ and x_2 the vertex $r'_p(x) \in V_{R_p}$. As $\bar{G}$ is a structured hypergraph, x_1 is adjacent to one ○-vertex v_1 and one □-vertex e_1. Analogously, x_2 is adjacent to one ○-vertex v_2 and one □-vertex e_2 in $\bar{R}_p$. Then the S-graph morphisms φ_X and r_p must map e_1 and e_2 on the only □-vertex in $\bar{A}^I$ which is adjacent to x'. If, moreover, x' is adjacent to the ○-vertex v' in $\bar{A}^C$, then they map both v_1 and v_2 on v'. Otherwise, x' is adjacent to all ○-vertices in $\bar{A}^U$. Then the unknown part of $\bar{R}_p$ is not empty and mapped on the same k-copy of S in $\bar{A}^U$ as the unknown part of $\bar{G}$, which means that v_1 and v_2 have the same image. By the construction of a pullback, we now have for $\bar{H}$ that x is adjacent to a ○-vertex which is mapped to v_1 by φ'_u and to v_2 by r'_p, and to a □-vertex which is mapped to e_1 by φ'_X and to e_2 by r'_p. Finally, the fact that in $\bar{G}$ and $\bar{R}_p$ each •-vertex is adjacent to exactly one ○-vertex and exactly one □-vertex is sufficient to ensure that each •-vertex of $\bar{H}$ is adjacent to at most one ○-vertex and at most one □-vertex.

The interface part of $\bar{H}$ does not contain any ○-vertex, and in particular no nonterminal ○-vertex whose completion to a copy of S would have to be checked, so $\bar{H}$ is indeed a structured hypergraph. □

7.18 Theorem (correct translation)

Let $G \Rightarrow_{(v,p)} H$ be the application of a production p to a node v in a multiple hypergraph G and $\bar{G} \to_{(\varphi_X, r_p)} \bar{H}$ the corresponding pullback rule application. Then $\bar{H}$ is isomorphic to the expanded structured hypergraph associated with H.

Proof. As already stated in the proof of Lemma 7.17, the context part of $\bar{H}$ consists of an isomorphic copy of $\bar{G}$ minus the vertex v and its 'incident hyperedges,' and the unknown part of $\bar{H}$ consists of an isomorphic copy of $\bar{R}^U$, the structured hypergraph corresponding to the labelled hypergraph R in the production $p = (A ::= (R, C))$. Moreover, by Definition 7.15 these parts are expanded in the correct way.

Now we have to verify that the interface part of $\bar{H}$ contains exactly the embedding hyperedges created in the hNCE rewriting step, with the correct expansion. If there is an embedding hyperedge $(e, (ex/cr))$ in H, then the construction of φ_X and r_p ensures that the corresponding $\Box$-vertex e together with the correct $\bullet$-vertices belongs to $\bar{H}$ and is assigned the expansion triple $Ex(cr)$. Conversely, the pullback construction implies that a $\Box$-vertex e is in the interface part of $\bar{H}$ if and only if φ_X maps the $\Box$-vertex $\varphi'_X(e)$ in the interface part of $\bar{G}$ to the $\Box$-vertex e' in $\bar{A}^I$ and e' is also the image under r_p of the $\Box$-vertex $r'_p(e)$ in the interface part of $\bar{R}_p$. Then, the transformed expansion triple of $\hat{e} = \varphi'_X(e)$ with respect to v as nonterminal is some Ex, $r'_p(e)$ is the $\Box$-vertex of some S-graph $\lfloor Cr \rfloor$ in the interface part of $\bar{R}_p$ which is assigned the expansion triple $Ex(cr)$, and the connection instruction (ex/cr) is in C. Therefore, $(\hat{e}, (ex/cr))$ must be an embedding hyperedge in H, with label $lab(cr)$ and its ith incident node v_i labelled $ex(cr)[i]$, which implies that the expansion triple $Ex(cr)$ assigned to e is correct. Whether v_i lies in R or is the $cr[i]$th incident node of $\hat{e}$ in G, it is not difficult to see that its copy in the unknown resp. context part of $\bar{H}$ is the $\circ$-vertex adjacent to the ith $\bullet$-vertex adjacent to e. Thus, the interface part of $\bar{H}$ corresponds indeed to the embedding hyperedges of H. $\Box$

7.5   Pullback Rewriting with Net Morphisms

In this section, pullback rewriting is considered for the category of nets and net morphisms. Although net morphisms are similar to hypergraph morphisms, the respective categories differ notably in that the category of net morphisms is not complete. Nevertheless, pullback rewriting can be defined in a meaningful way, allowing to specify net refinements [Kle98]; indeed, pullback rewriting in nets has been related to pullback rewriting in hypergraphs in the same reference.

Instead of developing an abstract framework for any refinement operation, van Glabbeek and Goltz's transition refinement operation [GG90], also studied in Section 6.3, will be translated in terms of pullback rewriting. As the main part of this

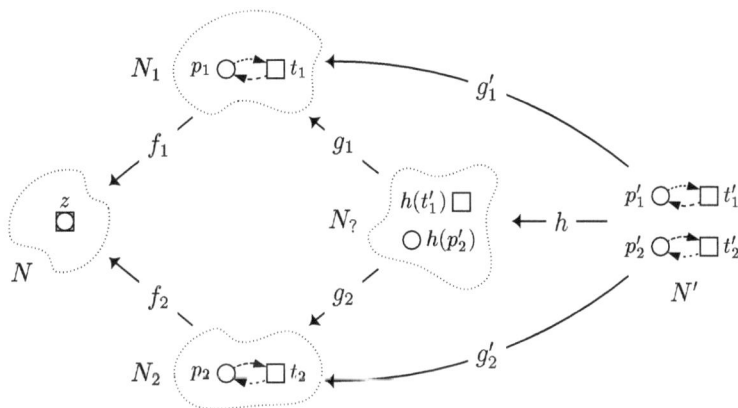

Figure 7.15: Illustrating the proof of Theorem 7.19

operation concerns the transformation of one net structure into another, we will concentrate on net structures and net morphisms, disregarding markings.

Net structures and net morphisms as defined in Sections 6.1 and 6.2 form a category which is denoted by $\mathcal{N}$. The terminal objects of $\mathcal{N}$ are $\bigcirc$ and $\square$. Unlike the categories of structured graphs, $\mathcal{N}$ is not complete; in particular, certain pullbacks do not exist. The pairs of net morphisms for which a pullback can be constructed are characterised as follows.

7.19 Theorem (characterisation of net pullbacks)
For $i \in \{1,2\}$, let $N_i = (P_i, T_i, F_i)$ and $N = (P, T, F)$ be net structures and $f_i : N_i \to N$ net morphisms. The pullback of (f_1, f_2) exists if and only if for every item $z \in P \cup T$ of N, at most one of the sets $f_1^{-1}(z)$, $f_2^{-1}(z)$ contains distinct items x and y such that the arc (x, y) belongs to the flow relation of the corresponding net structure.

Proof. '$\Rightarrow$': Let $z \in P \cup T$ and, for $i \in \{1, 2\}$, $p_i, t_i \in f_i^{-1}(z)$ with $(p_i, t_i) \in F_i$ or $(t_i, p_i) \in F_i$. Moreover, let $N_?$ be a net structure and $g_1 : N_? \to N_1$, $g_2 : N_? \to N_2$ net morphisms with $g_1 \circ f_1 = g_2 \circ f_2$. We show that no choice for $(N_?, g_1, g_2)$ can be a pullback of (f_1, f_2).

Let N' be the net structure with places p_1', p_2', transitions t_1', t_2', and, for each $i \in \{1,2\}$, an arc between p_i' and t_i' mirroring (one of) the arc(s) between p_i and t_i. Now consider the two net morphisms $g_1' : N' \to N_1$, $g_2' : N' \to N_2$ with $g_1'(p_1') = p_1$, $g_1'(\{t_1', p_2', t_2'\}) = \{t_1\}$, $g_2'(\{p_1', t_1', p_2'\}) = \{p_2\}$, and $g_2'(t_2') = t_2$; clearly, $g_1' \circ f_1 = g_2' \circ f_2$. Finally, let $h : N' \to N_?$ be a net morphism such that $g_1 \circ h = g_1'$ and $g_2 \circ h = g_2'$. This situation is depicted in Figure 7.15.

As $g_i \circ h = g_i'$ maps p_i' to p_i and t_i' to t_i, h does not identify p_i' and t_i', for each $i \in \{1,2\}$. Therefore, the arc between p_1' and t_1' resp. p_2' and t_2' implies that $h(t_1')$ is a transition and $h(p_2')$ a place. Moreover, $g_i \circ h = g_i'$ identifying t_1' and p_2' means that g_i identifies $h(t_1')$ and $h(p_2')$, for each $i \in \{1,2\}$. Hence, for net morphisms $g_1'': N' \rightarrow N_1$ with $g_1''(P' \cup T') = \{t_1\}$ and $g_2'': N' \rightarrow N_2$ with $g_2''(P' \cup T') = \{p_2\}$, the two distinct net morphisms $h_1, h_2: N' \rightarrow N_?$ with $h_1(P' \cup T') = \{h(t_1')\}$ and $h_2(P' \cup T') = \{h(p_2')\}$ satisfy $g_i \circ h_j = g_i''$, for all $i, j \in \{1,2\}$. Thus, (g_1, g_2) cannot be the pullback of (f_1, f_2).

'$\Leftarrow$': Let, for $i \in \{1,2\}$, Z_i be the set of items in N on which f_i maps an arc, i.e. $Z_i = \{z \in P \cup T \mid \exists x, y \in f_i^{-1}(z)$ with $(x,y) \in F_i\}$. By assumption, Z_1 and Z_2 are disjoint. Let N_{pb} be as follows:

- $P_{pb} = \{(x_1, x_2) \in f_1^{-1}(z) \times f_2^{-1}(z) \mid (z \in P \setminus (Z_1 \cup Z_2))$ or $(z \in Z_1$ and $x_1 \in P_1)$ or $(z \in Z_2$ and $x_2 \in P_2)\}$,

- $T_{pb} = \{(x_1, x_2) \in f_1^{-1}(z) \times f_2^{-1}(z) \mid (z \in T \setminus (Z_1 \cup Z_2))$ or $(z \in Z_1$ and $x_1 \in T_1)$ or $(z \in Z_2$ and $x_2 \in T_2)\}$,

- $F_{pb} = \{((x_1, x_2), (y_1, y_2)) \in (P_{pb} \times T_{pb}) \cup (T_{pb} \times P_{pb}) \mid (x_i, y_i) \in F_i$ and $(x_j = y_j$ or $(x_j, y_j) \in F_j)$ for $i, j \in \{1,2\}, i \neq j\}$.

Clearly, N_{pb} is a net structure. It is straightforward to verify that the projections $g_i: N_{pb} \rightarrow N_i$ with $g_i((x_1, x_2)) = x_i$ form the pullback of (f_1, f_2). □

In order to specify the refinement operation of [GG90] in terms of pullback rewriting, an alphabet structure, unknown morphisms, rule morphisms, and the application of rules to unknowns have to be defined. As the operation refines transitions, it suffices to consider an alphabet structure with an unknown part consisting of transitions only; in fact, exactly one unknown transition will be used. Moreover, the interface part of the refinement alphabet will consist entirely of places, and the context part of one transition. Figure 7.16 shows how the refinement depicted in Figure 6.10 can be converted into a pullback rewriting step.

7.20 Definition (pullback refinement)
The *refinement alphabet* is the net structure $N_A = (P_A, T_A, F_A)$ with $P_A = \{p_1, p_2\}$, $T_A = \{t_0, t_{-1}\}$, and $F_A = \{(t_0, p_1), (p_1, t_0), (t_0, p_2), (p_2, t_0), (p_1, t_{-1}), (t_{-1}, p_2)\}$.

Let N_1 be a net structure and t a transition in N_1. The *refinement unknown* on t is the net morphism $u_t: N_1 \rightarrow N_A$ with $u_t^{-1}(t_{-1}) = \{t\}$, $u_t^{-1}(p_1) = {}^\bullet t$, and $u_t^{-1}(p_2) = t^\bullet$.

Let N_2 be a net structure underlying some refinement net and N_2^+ the net structure obtained from N_2 by adding a new transition t^+, all possible arcs from t^+ to the initial and terminal places of N_2, and all possible arcs from these places to

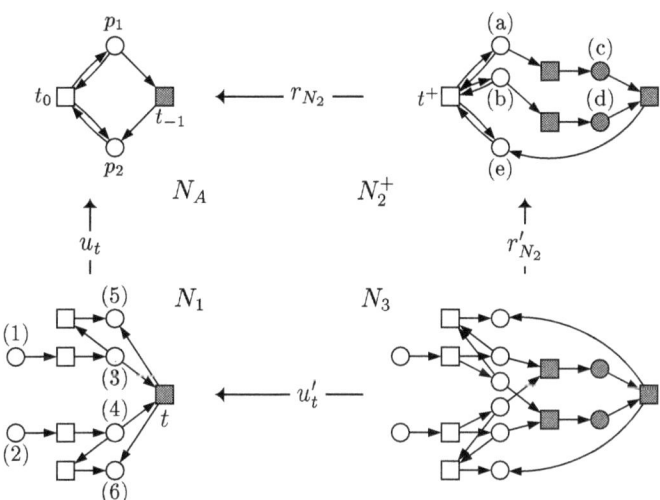

Figure 7.16: Transition refinement as pullback rewriting

t^+. The *refinement rule* induced by N_2 is the net morphism $r_{N_2} \colon N_2^+ \to N_A$ with $r_{N_2}^{-1}(t_0) = \{t^+\}$, $r_{N_2}^{-1}(p_1) = {}^\circ N_2$, and $r_{N_2}^{-1}(p_2) = N_2^\circ$.

Let $u_t \colon N_1 \to N_A$ be a refinement unknown and r_{N_2} a refinement rule. The *application* of r_{N_2} to u_t is obtained by computing the pullback of u_t and r_{N_2}, and we write $N_1 \to_{(u_t, r_{N_2})} N_3$, where N_3 is the pullback object.

Note that the flow relation of N_A is not symmetric, leading to a natural separation of the preplaces and the postplaces of any transition to be refined. As all nets are assumed to be loop-free, a refinement unknown can be defined on any transition, and is unique in every case.

With pullback refinement defined as above, Theorem 7.19 allows to state that a pullback can be constructed for every pair of refinement unknown and refinement rule.

7.21 Corollary (well-definedness of pullback refinement)
For every refinement unknown u_t and every refinement rule r_N, the application of r_N at u_t is defined.

Proof. Of N_A, only the item t_0 (resp. t_{-1}) may contain an arc in its inverse image under u_t (resp. under r_N). As $t_0 \neq t_{-1}$, Theorem 7.19 implies the assertion.   □

Moreover, this notion of pullback refinement is indeed an implementation of the net refinement operation from [GG90].

7.22 Theorem (correctness of pullback refinement)

Let N_1 be a net structure containing a transition t, and let N_2 be a net structure underlying some refinement net. Moreover, let $N_1 \to_{(u_t, r_{N_2})} N_3$, where u_t is the refinement unknown on t, and r_{N_2} the refinement rule induced by N_2. Then N_3 and $N_1[t/N_2]$ are isomorphic.

Proof. By construction, N_3 and $N_1[t/N_2]$ only differ in that N_3 contains an item (x, t^+) for each $x \in (P_1 \cup T_1) \smallsetminus (^\bullet t \cup \{t\} \cup t^\bullet)$ and an item (t, y) for each $y \in (P_2 \smallsetminus (^\circ N_2 \cup N_2^\circ)) \cup T_2$. □

Finally, note that the morphism $u'_t \colon N_3 \to N_1$ generated by the pullback construction specifies the transition refinement in the sense of Definition 6.2. In fact, it is a vicinity-respecting morphism in the sense of [DM91], see [GG90].

7.6   Concluding Remarks

In Section 7.4, see also [JK00], we have shown that hNCE rewriting can be correctly translated into the pullback approach. Thus, it is as reasonable to consider the pullback approach as a generic framework for node rewriting as it is to consider the hNCE approach for node rewriting in directed labelled hypergraphs.

In the translation, the interface parts of the structured hypergraphs involved in the encoding are chosen such that they do not contain ∘-vertices. This shows that the embedding process of the hNCE approach does not need to have the information on the labels of the nodes adjacent to the rewritten one. For the translation into the pullback approach, however, there is an alternative, see Figure 7.17 which takes up the example from the previous section: The interface part of the structured alphabet graph as well as of the domain of a rule can also contain one ∘-vertex for

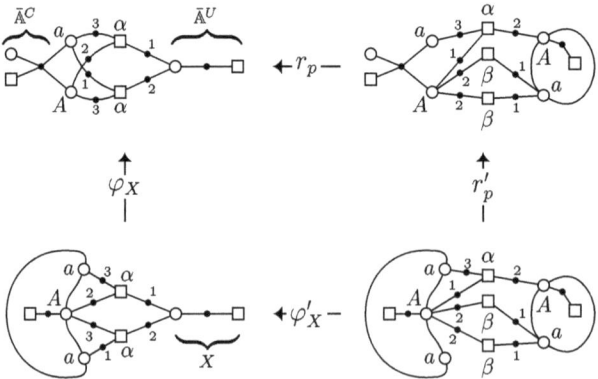

Figure 7.17: An alternative encoding of an hNCE rewriting step

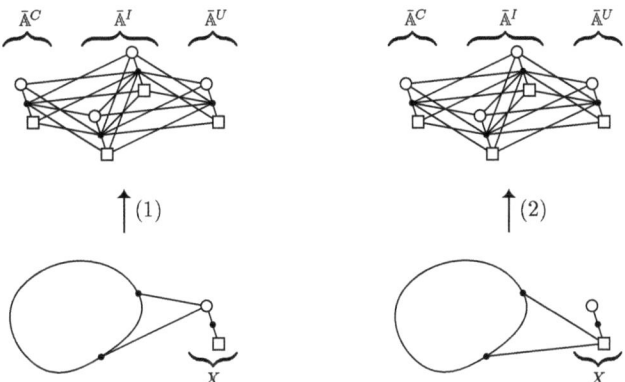

Figure 7.18: A 'node rewriting' (1) and a 'hyperedge rewriting' unknown (2)

each (node) label, just as in Bauderon's (graph-rewriting) NCE encoding [Bau96]. Note that this alternative results in a partition of the interface parts into a section which remains unchanged during a rewriting step—the nodes—and a section where changes may occur—the linking hyperedges.

The translation is insofar a sketch as there is no explanation just how the role of nonterminal node labels is filled. In [BJ01b, Definition 14], an S-graph is 'labelled' with a set of unknown morphisms, where the set contains for every nonterminal copy X of S exactly one unknown on X. Thus, a nonterminal node label is translated into one type of unknown morphisms, i.e. it corresponds to one of the copies of S in the unknown part of the alphabet. This in turn implies that if two hNCE productions have the same left-hand side, then their pullback counterparts must have the same type. An alternative, which abstracts from the notion of a nonterminal node label, is to have, for every k for which there is a k-copy of S in $\bar{\mathbb{A}}^U$, exactly one pullback rule of type k. Then for every nonterminal copy X of S in an S-graph, there is a set of unknowns on X of pairwise distinct type, and identical behaviour otherwise. The nondeterminism of the classical rewriting process finds its expression by choosing first some nonterminal copy X of S, and then one of the unknowns on X (which immediately determines the rule to be applied).

As Bauderon and Jacquet show in [BJ01a, Section 4] (for the category of graphs structured by $\circ\!\!-\!\!\square$), pullback rewriting also allows to encode hyperedge rewriting in directed hypergraphs. In fact, the generic form of the structured alphabet $\bar{\mathbb{A}}_S$ is the same for both node and hyperedge rewriting. Only the way in which the unknown part of the domain of an unknown morphism is linked to the interface part allows to distinguish whether a hyperedge or a node is being rewritten; see the sketches in Figure 7.18. Thus, the pullback approach can be investigated as a natural formal basis for a combination of node and hyperedge rewriting. It should be noted, however, that the notion of context-freeness has as yet not been considered for pullback rewriting grammars.

Turning to the category-theoretical treatment of net refinement, we have seen at the example of the operation from [GG90] that pullback rewriting with net morphisms provides an elegant formalisation. The Petri net operator proposed in [Vog89] for the parallel composition with synchronisation is another likely candidate to be implemented in this framework. Furthermore, it should be interesting to consider the vicinity-respecting net morphisms of [DM91] for pullback rewriting. In addition to the refinement example of [GG90], [Kle98] contains a more general implementation of place refinement, which is moreover related to the pullback implementation of node rewriting in the category of directed graphs structured by ⊙ studied by Jacquet [Jac99, Chapter 6].

Conclusion

Summary

In the present thesis, context-free hypergraph rewriting is studied from various perspectives.

The technique of hNCE rewriting is introduced as a node-rewriting approach for hypergraph transformation. Confluent hNCE grammars are context-free, and their generative power is established to be properly superior to that of the two major approaches to context-free hypergraph rewriting, namely hyperedge-rewriting and separated handle-rewriting grammars. This relationship holds already for remote-free C-hNCE grammars, whose graph-generating power is nevertheless equal to that of the well-known C-edNCE graph grammars. Thus, C-hNCE rewriting is an important addition to the known context-free hypergraph-rewriting approaches. Moreover, it retains sufficient similarities to C-edNCE rewriting so that results on limitations of the generative power can be transferred.

Considering on the one hand the importance of confluent hNCE grammars and on the other that confluence is a dynamic property of the derivations in an hNCE grammar, the question arises whether this property can be decided. Due to the high flexibility of the hNCE mechanism to create embedding hyperedges, the problem is considerably more complex than for edNCE grammars. Still, it is decidable, with time complexity at most doubly exponential in the size of an input grammar.

The idea motivating this thesis was to combine node and hyperedge rewriting into one approach—called atom replacement—for context-free hypergraph rewriting. The fact that hyperedge rewriting can be encoded into hNCE rewriting by a one-to-one simulation of derivation steps seems at first to contradict the usefulness of such a combination. However, atom replacement allows to obtain in an intuitive way a number of graph and hypergraph languages which are not C-hNCE languages. While the grammars in question are associative and confluent, they are not context-free because they do not preserve (the number of) nonterminals. In fact, the languages are basically generated by first using a confluent hNCE grammar producing hypergraphs with a high number of hyperedges, and subsequently replacing these hyperedges. This technique is reflected in the closure of atom-replacement grammars under hyperedge substitution.

Interpreting hypergraphs as Petri nets, context-free hypergraph rewriting approaches provide models for the refinement of transitions (and places). While some refinement methods can be captured by hyperedge rewriting, others need the hyperedge multiplication possible in node rewriting. Given a one-to-one correspondence between rewriting steps and refinement steps, a context-free hypergraph grammar formalises restrictions to the structure of admissible Petri nets in the same way that a context-free string grammar is used to select strings in the correct format for a program in some programming language. Thus, sets of Petri nets can be specified which are suitable for some particular application domain such as e.g. models of workflow processes.

Thinking in the language of category theory, the selective multiplication of hyperedges occurring during a node-rewriting step is expressed in the notion of a fibred product, i.e. a pullback. Based on this idea, hNCE rewriting can be translated into terms of pullback rewriting. The same is true for certain Petri net refinements, with the added advantage that one can circumvent the encoding of Petri nets as hypergraphs and make direct use of net morphisms instead.

Outlook

The research presented here can be continued in a number of directions.

Node rewriting in hypergraphs. The material on C-hNCE grammars has been selected to concentrate on questions and problems which are germane to this approach, in particular to dealing with hyperedges instead of edges. For instance, the search for a simpler process to create embedding hyperedges led to the link-preserving and form-preserving normal forms studied in Chapter 3 resp. discussed in Section 4.4, and one may well think of further simplifications. Still, it turned out that several results can be generalised from the graph to the hypergraph case (and put to good use there). Therefore, a similar stability may be expected for other results on C-edNCE rewriting. This should nevertheless be verified in a systematic way, for which the survey in [ER97] offers an excellent pattern. For example, generalising static confluence to hNCE rewriting and proving its equivalence to dynamic confluence with respect to expressive power (cf. [ER97, Proposition 1.3.6]) would allow a significantly more comfortable (static) confluence check than the test developed in Chapter 4. A second example is provided by the proof of [ER97, Theorem 1.3.11] where for any edNCE grammar NG, a C-edNCE grammar NG' is constructed such that $L(NG')$ coincides with the language of graphs generated by leftmost derivations in NG. In conjunction with a simple bottom-up transduction of the derivation trees in a C-hNCE grammar, the generalisation of this construction would imply that every C-hNCE grammar has a remote-free normal form.

As a mathematical language to treat the hypergraph grammars in this thesis, mainly set theory has been used, with an excursion to category theory in Chapter 7. Arguably the most important alternative to express graph transformations employs monadic second-order logic, see [Cou97]. In order to place C-hNCE rewriting in this framework, hypergraphs are understood as relational structures in the sense of [Cou92]. For a translation of the actual rewriting mechanism into operations on relational structures, it may be a good idea to start out with 'simple' connection instructions, e.g. form-preserving ones, and then work towards the general format. The aim is a logical characterisation of C-hNCE languages, which would probably allow to obtain alternative proofs for the results of Chapter 3, definitely entail attractive closure properties, and generally lead to deeper insights into the nature of context-free node rewriting in hypergraphs.

Combining node and hyperedge rewriting. In its most general form, atom replacement allows to insert embedding hyperedges and to identify nodes independently of the type—node or hyperedge—of the replaced item. Imposing context-freeness, one may expect that hyperedge-rewriting steps can be simulated by node-rewriting steps. In turn, this implies that context-free atom-replacement grammars can generate more hypergraph languages than C-hNCE grammars if and only if the difference is a consequence of the possibility to glue, in a node-rewriting step, nodes of the replacing hypergraph with neighbours of the replaced node.

On the other hand, confluent basic atom-replacement grammars, which combine ordinary node-rewriting steps with ordinary hyperedge-rewriting steps, are 'quasi-context-free' in the sense that they are associative and confluent, but not nonterminal-preserving. These properties lead to a notion of derivation trees where a vertex has a fixed *minimal* number of children, which is the number of nonterminal items in the right-hand side hypergraph of the production labelling that vertex. If the rewritten item is a hyperedge, that is the absolute number of children; otherwise, there may be additional children due to nonterminal embedding hyperedges. It may be possible to find, at least for certain cases, a formula which describes the number of these additional children depending on the depth of the parent node in the derivation tree. This might then lead to a variation of Parikh's Theorem, allowing to prove for some hypergraph languages that they cannot be generated with C-BAR grammars. Moreover, the derivation trees may be helpful to determine a characterisation (of a subset) of C-BAR languages as iterations of hyperedge substitutions with node-rewriting languages.

Pullback rewriting. Using the language of category theory, the pullback rewriting approach offers a very natural, elegant, and—once one is used to it—intuitive formalisation of node rewriting. As it offers a generic framework for various graph- and hypergraph-rewriting approaches, it would be interesting to have generic conditions ensuring context-freeness. It seems possible that pullback rewriting is naturally associative. Confluence is guaranteed whenever the boundary condition (introduced in [RW86] for NLC grammars) is satisfied; more general conditions may be thought of. In order to verify nonterminal preservation, however, one first needs a generic definition of the reconstruction of unknowns after a rewriting step.

As hNCE rewriting can simulate hyperedge rewriting and be translated into pullback rewriting, so can hyperedge rewriting be expressed by pullbacks. Hence, it may well be worthwhile to study a combination of node and hyperedge rewriting in hypergraphs which is based on the pullback mechanism. This may possibly be combined with a further investigation of the parallelisation properties associated with pullback rewriting, so that nodes and hyperedges can be rewritten in parallel with the construction of just one pullback square. Indeed, there may even be a link to hyperedge rewriting with rendezvous [DDK93], a technique where hyperedges are rewritten in parallel and nodes of distinct replacing hypergraphs may be glued with each other provided the replaced hyperedges shared some incident node(s). In contrast to this, a simulation of general atom replacement cannot be expected because it permits gluing two distinct nodes in the remainder hypergraph (by gluing them with the same node in the replacing hypergraph), which amounts to identifying distinct interface nodes and cannot be modelled in a pullback.

Application to Petri nets. Many Petri net refinement operations proposed in the literature have a context-free nature. For a selected few refinement techniques, this has been made precise in Chapter 6 by implementing them as context-free hypergraph rewriting. While implementations for other refinement mechanisms can be sought, it may be more exciting to exploit that context-freeness and explore further links between Petri net and hypergraph transformation, such as the correspondence between Petri net reductions and hypergraph parsing or between hierarchical Petri nets and hypergraph derivation trees.

Considering the more complex workflow net transformation rules mentioned in [vdA97] which are intended to model the reengineering of a workflow process—in that reference, a sample rule is given which has on one side the right part of rule (a) in Figure 6.16 and on the other the right part of rule (c)—corresponds, on the basis of a context-free grammar, to replacing one branch of a derivation tree with another. Given that the basic workflow refinement rules can be implemented as hyperedge rewriting, such a replacement rule can be formalised as a rule in the double-pushout approach [CMR$^+$97]. Analogously, if the underlying context-free rules are node-rewriting, it should be possible to express the replacement by a double-pullback rule [BJ01a]. Thinking along these lines, a general theory relating replacements on derivation trees with their effects on the generated structures should be made available for modelling and reengineering systems with Petri nets.

Bibliography

[AEH⁺99]   Marc Andries, Gregor Engels, Annegret Habel, Berthold Hoffmann, Hans-Jörg Kreowski, Sabine Kuske, Detlef Plump, Andy Schürr, and Gabriele Taentzer. Graph transformation for specification and programming. *Science of Computer Programming*, 34(1):1–54, 1999.

[AL91]   Andrea Asperti and Giuseppe Longo. *Categories, Types, and Structures*. The MIT Press, 1991.

[Bar99]   Klaus Barthelmann. *Über die Bedeutung der Graphersetzung für die Programmverifikation*. Habilitation thesis, Mainz, 1999.

[Bau95a]   Michel Bauderon. A uniform approach to graph rewriting: The pullback approach. In M. Nagl, editor, *Graph-Theoretic Concepts in Computer Science*, volume 1017 of *Lecture Notes in Computer Science*, pages 101–115. Springer, 1995.

[Bau95b]   Michel Bauderon. Parallel rewriting of graphs through the pullback approach. In [CM95], 8 pages.

[Bau96]   Michel Bauderon. A category-theoretical approach to vertex replacement: The generation of infinite graphs. In [CEER96], pages 27–37.

[BC87]   Michel Bauderon and Bruno Courcelle. Graph expressions and graph rewriting. *Mathematical Systems Theory*, 20:83–127, 1987.

[Ber73]   Claude Berge. *Graphs and Hypergraphs*. North-Holland, Amsterdam, 1973.

[Ber91]   Claude Berge. *Graphs*. Elsevier, Amsterdam, 1991.

[BGV91]   Wilfried Brauer, Robert Gold, and Walter Vogler. A survey of behaviour and equivalence preserving refinements of Petri nets. In G. Rozenberg, editor, *Advances in Petri Nets 1990*, volume 483 of *Lecture Notes in Computer Science*, pages 1–46. Springer, 1991.

[BJ01a]   Michel Bauderon and Hélène Jacquet. Pullback as a generic graph
          rewriting mechanism. *Applied Categorical Structures*, 9(1):65–82, 2001.

[BJ01b]   Michel Bauderon and Hélène Jacquet. Node rewriting in graphs and
          hypergraphs: a categorical framework. *Theoretical Computer Science*,
          266:463–487, 2001.

[BN98]   Franz Baader and Tobias Nipkow. *Term Rewriting and All That*. Cam-
          bridge University Press, 1998.

[Bol98]   Béla Bollobás. *Modern Graph Theory*, volume 184 of *Graduate Texts
          in Mathematics*. Springer, 1998.

[CEER96] Janice E. Cuny, Hartmut Ehrig, Gregor Engels, and Grzegorz Rozen-
          berg, editors. *Graph Grammars and Their Application to Computer
          Science*, volume 1073 of *Lecture Notes in Computer Science*. Springer,
          1996.

[CER79]   Volker Claus, Hartmut Ehrig, and Grzegorz Rozenberg, editors. *Graph
          Grammars and Their Application to Computer Science and Biology*,
          volume 73 of *Lecture Notes in Computer Science*. Springer, 1979.

[CER93]   Bruno Courcelle, Joost Engelfriet, and Grzegorz Rozenberg. Handle-
          rewriting hypergraph grammars. *Journal of Computer and System Sci-
          ences*, 46:218–270, 1993.

[Che76]   Wai-Kai Chen. *Applied Graph Theory. Graphs and Electrical Networks*.
          North-Holland, Amsterdam, 1976.

[CM95]   Andrea Corradini and Ugo Montanari, editors. *Proc. Joint COMPU-
          GRAPH/SEMAGRAPH Workshop on Graph Rewriting and Compu-
          tation (SEGRAGRA'95)*, volume 2 of *Electronic Notes in Theoretical
          Computer Science*. Elsevier, 1995.

[CMR⁺97] Andrea Corradini, Ugo Montanari, Francesca Rossi, Hartmut Ehrig,
          Reiko Heckel, and Michael Löwe. Algebraic approaches to graph trans-
          formation - part I: Basic concepts and double pushout approach. In
          [Roz97], chapter 3, pages 163–246.

[Cou87]   Bruno Courcelle. An axiomatic definition of context-free rewriting and
          its application to NLC graph grammars. *Theoretical Computer Science*,
          55:141–181, 1987.

[Cou92]   Bruno Courcelle. The monadic second-order logic of graphs VII: Graphs
          as relational structures. *Theoretical Computer Science*, 101:3–33, 1992.

[Cou97]   Bruno Courcelle. The expression of graph properties and graph trans-
           formations in monadic second-order logic. In [Roz97], chapter 5, pages
           313–400.

[DDK93]   Gnanamalar David, Frank Drewes, and Hans-Jörg Kreowski. Hyper-
           edge replacement with rendezvous. In P. Jouannaud, editor, *Theory
           and Practice of Software Development*, volume 668 of *Lecture Notes in
           Computer Science*, pages 167–181, 1993.

[DHK97]   Frank Drewes, Annegret Habel, and Hans-Jörg Kreowski. Hyperedge
           replacement graph grammars. In [Roz97], chapter 2, pages 95–162.

[Dic93]   Emily Dickinson. *Selected Poems*. Gramercy Books, New York · Avenel,
           New Jersey, 1993.

[Die00]   Reinhard Diestel. *Graph Theory*, volume 173 of *Graduate Texts in
           Mathematics*. Springer, 2nd edition, 2000.

[Dij71]   E.W. Dijkstra. Hierarchical ordering of sequential processes. *Acta In-
           formatica*, 1:115–138, 1971.

[DM91]   Jörg Desel and Agathe Merceron. Vicinity respecting net morphisms.
           In G. Rozenberg, editor, *Advances in Petri Nets 1990*, volume 483 of
           Lecture Notes in Computer Science, pages 165–185. Springer, 1991.

[EEKR99]   Hartmut Ehrig, Gregor Engels, Hans-Jörg Kreowski, and Grzegorz
           Rozenberg, editors. *Handbook of Graph Grammars and Computing by
           Graph Transformation*, volume 2. World Scientific, 1999.

[EEKR00]   Hartmut Ehrig, Gregor Engels, Hans-Jörg Kreowski, and Grzegorz
           Rozenberg, editors. *Theory and Application of Graph Transformations*,
           volume 1764 of *Lecture Notes in Computer Science*. Springer, 2000.

[Ehr79]   Hartmut Ehrig. Introduction to the algebraic theory of graph gram-
           mars. In [CER79], pages 1–69.

[EKMR99]   Hartmut Ehrig, Hans-Jörg Kreowski, Ugo Montanari, and Grzegorz
           Rozenberg, editors. *Handbook of Graph Grammars and Computing by
           Graph Transformation*, volume 3. World Scientific, 1999.

[EKR91]   Hartmut Ehrig, Hans-Jörg Kreowski, and Grzegorz Rozenberg, editors.
           Graph Grammars and Their Application to Computer Science, volume
           532 of *Lecture Notes in Computer Science*. Springer, 1991.

[EL89]   Joost Engelfriet and George Leih. Linear graph grammars: Power and
           complexity. *Information and Computation*, 81:88–121, 1989.

[Eng97]   Joost Engelfriet. Context-free graph grammars. In G. Rozenberg and
           A. Salomaa, editors, *Handbook of Formal Languages. Vol. III: Beyond
           Words*, chapter 3, pages 125–213. Springer, 1997.

[ENR83]   Hartmut Ehrig, Manfred Nagl, and Grzegorz Rozenberg, editors. *Graph
           Grammars and Their Application to Computer Science*, volume 153 of
           Lecture Notes in Computer Science. Springer, 1983.

[ENRR87]   Hartmut Ehrig, Manfred Nagl, Grzegorz Rozenberg, and Azriel Rosen-
           feld, editors. *Graph Grammars and Their Application to Computer
           Science*, volume 291 of *Lecture Notes in Computer Science*. Springer,
           1987.

[EPS73]   Hartmut Ehrig, M. Pfender, and H. J. Schneider. Graph grammars: An
           algebraic approach. In *IEEE Conf. on Automata and Switching Theory*,
           pages 167–180, Iowa City, 1973.

[ER90]   Joost Engelfriet and Grzegorz Rozenberg. A comparison of boundary
           graph grammars and context-free hypergraph grammars. *Information
           and Computation*, 84:163–206, 1990.

[ER97]   Joost Engelfriet and Grzegorz Rozenberg. Node replacement graph
           grammars. In [Roz97], chapter 1, pages 1–94.

[Fed71]   Jerome Feder. Plex languages. *Information Sciences*, 3:225–241, 1971.

[Feh93]   Rainer Fehling. A concept of hierarchical Petri nets with building
           blocks. In G. Rozenberg, editor, *Advances in Petri Nets 1993*, volume
           674 of *Lecture Notes in Computer Science*, pages 148–168. Springer,
           1993.

[GG90]   Rob van Glabbeek and Ursula Goltz. Refinement of actions in causal-
           ity based models. In J.W. de Bakker, W.-P. de Roever, and G. Rozen-
           berg, editors, *Stepwise Refinement of Distributed Systems. Models, For-
           malisms, Correctness*, volume 430 of *Lecture Notes in Computer Sci-
           ence*, pages 267–300. Springer, 1990.

[GSW80]   H.J. Genrich and E. Stankiewicz-Wiechno. A dictionary of some basic
           notions of net theory. In W. Brauer, editor, *Net Theory and Applica-
           tions*, volume 84 of *Lecture Notes in Computer Science*, pages 519–535.
           Springer, 1980.

[Hab92a]   Annegret Habel. *Hyperedge Replacement: Grammars and Languages*, volume 643 of *Lecture Notes in Computer Science*. Springer, 1992.

[Hab92b]   Annegret Habel. Hypergraph grammars: Transformational and algorithmic aspects. *Journal of Information Processing and Cybernetics EIK*, 28:241–277, 1992.

[Har69]   Frank Harary. *Graph Theory*. Addison-Wesley, Reading, Mass., 1969.

[HK87]   Annegret Habel and Hans-Jörg Kreowski. May we introduce to you: hyperedge replacement. In [ENRR87], pages 15–26.

[HK98]   Annegret Habel and Renate Klempien-Hinrichs. Atom replacement in hypergraphs. In G. Engels and G. Rozenberg, editors, *Preproc. TAGT'98*, Bericht tr-ri-98-201, Reihe Informatik, pages 182–189. Universität–GH Paderborn, 1998.

[HS79]   Horst Herrlich and George Strecker. *Category Theory*. Heldermann, Berlin, second edition, 1979.

[HU79]   John E. Hopcroft and Jeffrey D. Ullman. *Introduction to Automata Theory, Languages, and Computation*. Addison-Wesley, Reading, Mass., 1979.

[Jac99]   Hélène Jacquet. *Une approche catégorique de la réécriture de sommets dans les graphes*. Doctoral thesis, Université Bordeaux I, 1999.

[JK00]   Hélène Jacquet and Renate Klempien-Hinrichs. Node replacement in hypergraphs: Translating NCE rewriting into the pullback approach. In [EEKR00], pages 117–130.

[JKRE82]   Dirk Janssens, Hans-Jörg Kreowski, Grzegorz Rozenberg, and Hartmut Ehrig. Concurrency of node-label-controlled graph transformations. In H.-J. Schneider and H. Göttler, editors, *Proc. 8th Conference on Graph-theoretic Concepts in Computer Science (WG'82)*. Carl Hauser Verlag, München Wien, 1982.

[Kau85]   Manfred Kaul. *Syntaxanalyse von Graphen durch Präzedenzgraphgrammatiken*. Doctoral thesis, Universität Osnabrück, 1985.

[KJ99]   Changwook Kim and Tae Eui Jeong. HRNCE grammars – a hypergraph generating system with an eNCE way of rewriting. *Theoretical Computer Science*, 233:143–178, 1999.

[KK99]   Hans-Jörg Kreowski and Sabine Kuske. Graph transformation units
            and modules. In [EEKR99], pages 607–638.

[Kle96]   Renate Klempien-Hinrichs. Node replacement in hypergraphs: Sim-
            ulation of hyperedge replacement, and decidability of confluence. In
            [CEER96], pages 397–411.

[Kle98]   Renate Klempien-Hinrichs. Net refinement by pullback rewriting. In
            M. Nivat, editor, *Proc. Foundations of Software Science and Compu-
            tation Structures*, volume 1378 of *Lecture Notes in Computer Science*,
            pages 189–202. Springer, 1998.

[Kle99]   Renate Klempien-Hinrichs. The generative power of context-free node
            rewriting in hypergraphs. *Grammars*, 2:211–221, 1999.

[Kle]   Renate Klempien-Hinrichs. Normal forms for context-free node-
            rewriting hypergraph grammars. *Mathematical Structures in Computer
            Science*. To appear.

[KR90]   Hans-Jörg Kreowski and Grzegorz Rozenberg. On structured graph
            grammars, parts I and II. *Information Sciences*, 52:185–210 and 221–
            246, 1990.

[Kre81]   Hans-Jörg Kreowski. A comparison between Petri-nets and graph gram-
            mars. In H. Noltemeier, editor, *Proc. Intl. Workshop on Graphtheoric
            Concepts in Computer Science (WG'80)*, volume 100 of *Lecture Notes
            in Computer Science*, pages 306–317. Springer, 1981.

[Kus00a]   Sabine Kuske. More about control conditions for transformation units.
            In [EEKR00], pages 323–337.

[Kus00b]   Sabine Kuske. *Transformation Units—a Structuring Principle for
            Graph Transformation Systems*. Doctoral thesis, Universität Bremen,
            2000.

[MK80]   T. Murata and J.Y. Koh. Reduction and expansion of live and safe
            marked graphs. *IEEE Transactions on Circuits and Systems*, CAS-
            27(1):68–70, 1980.

[Mur89]   Tadao Murata. Petri nets: Properties, analysis and applications. *Pro-
            ceedings of the IEEE*, 77(4):541–580, 1989.

[Nag76]   Manfred Nagl. Formal languages of labelled graphs. *Computing*, 16:113–
            137, 1976.

[Nag79]   Manfred Nagl. *Graph-Grammatiken: Theorie, Anwendungen, Imple-mentierungen.* Vieweg, Braunschweig, 1979.

[Par66]   Rohit J. Parikh. On context-free languages. *Journal of the Association for Computing Machinery,* 13:570–581, 1966.

[Pav72]   Theodosios Pavlidis. Linear and context-free graph grammars. *Journal of the Association for Computing Machinery,* 19(1):11–23, 1972.

[PR69]   John L. Pfaltz and Azriel Rosenfeld. Web grammars. In *Intl. Joint Conference on Artificial Intelligence,* pages 609–619, 1969.

[Rei85]   Wolfgang Reisig. *Petri Nets. An Introduction,* volume 4 of *EATCS Monographs on Theoretical Computer Science.* Springer, 1985.

[Rei87]   Wolfgang Reisig. Petri nets in software engineering. In W. Brauer, W. Reisig, and G. Rozenberg, editors, *Petri Nets: Applications and Relationships to Other Models of Concurrency. Advances in Petri Nets 1986, Part II,* volume 255 of *Lecture Notes in Computer Science,* pages 63–96. Springer, 1987.

[Ros96]   Lexa Roséan. *The Supermarket Sorceress.* St. Martin's Press, New York, N.Y., 1996.

[Roz97]   Grzegorz Rozenberg, editor. *Handbook of Graph Grammars and Computing by Graph Transformation,* volume 1. World Scientific, 1997.

[RW86]   Grzegorz Rozenberg and Emo Welzl. Boundary NLC graph grammars—basic definitions, normal forms, and complexity. *Information and Control,* 69:136–167, 1986.

[Sal73]   Arto Salomaa. *Formal Languages.* Academic Press, New York, 1973.

[Sch70]   Hans Jürgen Schneider. Chomsky-Systeme für partielle Ordnungen. Arbeitsbericht 3,3, Institut für Mathematische Maschinen und Datenverarbeitung, Erlangen, 1970.

[SM83]   Ichiro Suzuki and Tadao Murata. A method for stepwise refinement and abstraction of Petri nets. *Journal of Computer and System Sciences,* 27:51–76, 1983.

[SW95]   Konstantin Skodinis and Egon Wanke. Emptiness problems of eNCE graph languages. *Journal of Computer and System Sciences,* 51:472–485, 1995.

[Val79]   R. Valette. Analysis of Petri nets by stepwise refinements. *Journal of Computer and System Sciences*, 18(1):35–46, 1979.

[vdA97]   W.M.P. van der Aalst. Verification of workflow nets. In P. Azéma and G. Balbo, editors, *Application and Theory of Petri Nets 1997 (Proc. ICATPN'97)*, volume 1248 of *Lecture Notes in Computer Science*, pages 407–426. Springer, 1997.

[vdA98]   W.M.P. van der Aalst. The application of Petri nets to workflow management. *Journal of Circuits, Systems and Computers*, 8(1):21–66, 1998.

[Vog87]   Walter Vogler. Behaviour preserving refinements of Petri nets. In G. Tinhofer and G. Schmidt, editors, *Graph-Theoretic Concepts in Computer Science (Proc. WG'86)*, volume 246 of *Lecture Notes in Computer Science*, pages 82–93. Springer, 1987.

[Vog89]   Walter Vogler. Failures semantics and deadlocking of modular Petri nets. *Acta Informatica*, 26:333–348, 1989.

[Vog91]   Walter Vogler. Failures semantics based on interval semiwords is a congruence for refinement. *Distributed Computing*, 4(3):139–162, 1991.

[Vog93]   Walter Vogler. Bisimulation and action refinement. *Theoretical Computer Science*, 114:173–200, 1993.

[Vog97]   Walter Vogler. A short story on action refinement. In C. Freksa, M. Jantzen, and R. Valk, editors, *Foundations of Computer Science. Potential – Theory – Cognition*, volume 1337 of *Lecture Notes in Computer Science*, pages 271–278. Springer, 1997.

[Wel84]   Emo Welzl. Encoding graphs by derivations and implications for the theory of graph grammars. In J. Paredaens, editor, *Proc. ICALP'84*, volume 172 of *Lecture Notes in Computer Science*, pages 503–513. Springer, 1984.

[Zub98]   W.M. Zuberek. Hierarchical derivation of schedules for manufacturing cells. In *Proc. IFAC Symp. on Information Control Problems in Manufacturing (INCOM'98)*, volume 2, pages 423–428, 1998.

List of Symbols

(continued from previous page)

(continued from previous page)

Index

I stepped from plank to plank
  So slow and cautiously;
The stars about my head I felt,
  About my feet the sea.

I knew not but the next
  Would be my final inch,—
This gave me that precarious gait
  Some call experience.

— Emily Dickinson —